Date Due

Phytochemical Ecology

Annual Proceedings of the Phytochemical Society

Phytochemical Ecology

PROCEEDINGS OF THE
PHYTOCHEMICAL SOCIETY SYMPOSIUM
ROYAL HOLLOWAY COLLEGE
ENGLEFIELD GREEN, SURREY
APRIL 1971

Edited by

J. B. HARBORNE

Department of Botany
The University of Reading, Reading, England

1972

ACADEMIC PRESS
LONDON AND NEW YORK

ACADEMIC PRESS INC. (LONDON) LTD.
24/28 Oval Road,
London NW1

United States Edition published by
ACADEMIC PRESS INC.
111 Fifth Avenue
New York, New York 10003

Library of Congress Catalog Card Number: 78-185208
ISBN: 0-12-324663-6

PRINTED IN GREAT BRITAIN BY
WILLIAM CLOWES & SONS LIMITED
LONDON, COLCHESTER AND BECCLES

Contributors

G. W. ARNOLD, *Division of Plant Industry, CSIRO, Floreat Park, Western Australia*

E. C. BATE-SMITH, *ARC Institute of Animal Physiology, Babraham, Cambridge, England*

E. A. BELL, *Department of Botany, University of Texas at Austin, Austin, Texas, U.S.A.*

J. M. CHERRETT, *Department of Applied Zoology, University College of North Wales, Bangor, Wales*

C. H. CHOU, *University of California, Santa Barbara, California, U.S.A.*

B. J. DEVERALL, *A. R. C. Unit on Plant Growth Substances and Systemic Fungicides, Wye College (University of London), Ashford, Kent, England*

W. G. H. EDWARDS, *Chemistry Department, Royal University of Malta, Malta*

J. L. HILL, *Division of Plant Industry, CSIRO, Floreat Park, Western Australia*

D. A. JONES, *Department of Genetics, The University, Birmingham, England*

A. R. MATTOCKS, *Toxicology Unit, Medical Research Council Laboratories, Woodmansterne Road, Carshalton, Surrey, England*

M. O. MOSS, *Department of Biological Sciences, University of Surrey, England*

C. H. MULLER, *University of California, Santa Barbara, California, U.S.A.*

T. A. ROHAN, *Bush Boake Allen Ltd., Hackney, London, England.*

M. ROTHSCHILD, *Ashton, Peterborough, Northamptonshire, England*

A. SHRIFT, *Department of Biological Sciences, State University of New York, Binghamton, New York, U.S.A.*

H. F. VAN EMDEN, *University of Reading, Horticultural Research Laboratories, Shinfield Grange, Shinfield, Berkshire, England*

Preface

While the so-called secondary metabolites of the plant kingdom—the alkaloids, cyanogenic glycosides, flavonoids and terpenoids—have always attracted scientists, particularly organic chemists, to the study of their structure, biosynthesis and natural distribution, much less attention has been given to their function. Indeed, in 1959, G. S. Fraenkel found it necessary to point out, in a now famous article in *Science*, that these substances do have a *raison d'être*, i.e. they are not just simply waste products of plant metabolism, as has so often been stated in the past. Since 1959, the tempo of research into the function of secondary constituents has steadily increased and we are at last in a position to assess the importance of these many and varied chemical structures in plant and animal ecology. The topic of phytochemical ecology— the biochemistry of plant and animal interactions—was therefore chosen for the major Annual Symposium of the Phytochemical Society of 1971, which was held in Royal Holloway College, London in April. The purpose of this volume is to present these Symposium papers to a wider audience, in the belief that they will be not only of interest to most research scientists but also that they will be a useful supplement to University teaching.

One of the best known examples of a complex chemical interaction between plants and animals is the case of the *Asclepias* cardiac glycosides, the Monarch butterfly and the Blue Jay. It is, therefore, entirely fitting that the first chapter in this volume should be by Miriam Rothschild, one of the authors of the *Asclepias* story, who presents here an account of her most recent experiments with aposematic butterflies and related insects. The relationship between insect feeding habits and the presence of repellents or attractants in plants is amply discussed in the next two chapters: J. M. Cherrett first considers chemical aspects of plant attack by leaf cutting ants and then H. F. van Emden covers chemical aspects of similar attack on crucifers and other plants by aphids.

The chemical interaction between plants and higher animals—a more complex situation than in the case of insects—is a subject about which few firm facts are available. However, as E. C. Bate-Smith mentions in Chapter 4, "much can be inferred from what is known about the chemistry of the plants that animals eat". Bate-Smith then refers to the condensed tannins, substances which produce the sensation of astringency, as being an apparently important group of feeding repellents in plants. As an example, he points to the fact that the presence of these tannins appears to determine which plant species are eaten or avoided by mountain gorillas in their native habitat in the Congo. The chemistry of flavour in foods is discussed in more detail in the next Chapter by T. A. Rohan, where the focus is on the interaction of man's own eating habits with secondary plant compounds. Rohan illustrates his theme with respect to the relationship between chemical structure and pungency, as found in ginger, pepper and capsicum. One of the problems of elucidating the chemical factors affecting selection of food plants by farm animals is that of communication; animals cannot talk. A method of overcoming this difficulty is to study animal preferences to solutions of pure chemicals and recent experiments on these lines are described in the following chapter by Arnold and Hill.

One of the main reasons why some plants are never eaten by animals is, of course, the occurrence of toxins, which are usually secondary constituents of one type or another. Ecological and biochemical aspects of plant toxins are appropriately covered in the next five chapters. D. A. Jones describes the cyanogenic glycosides which repel slugs and snails feeding on clover, M. O. Moss reviews the aflatoxin's and related mycotoxins, fungal metabolites produced in plants which can poison both turkeys and man, A. Shrift discusses organo-selenium derivatives of *Astragalus* and E. A. Bell covers the toxic amino acids of the Leguminosae and finally A. R. Mattocks deals with the *Senecio* alkaloids which are cattle poisons. In this last chapter, Mattocks discusses recent experiments which show that toxicity of these alkaloids is related to changes that occur during metabolism and that poisoning by pyrrolizidine alkaloids is due to the chemical interactions of reactive metabolites with vital cell constituents.

The last three chapters in the volume are devoted to chemical aspects of plant–plant interactions. The first by C. H. Muller and C. H. Chou describes the new areas of ecological research surrounding the phytotoxins—substances produced in one higher plant specifically for the purpose of preventing the growth of other plants in its vicinity. Another aspect of plant–plant interaction in which chemicals are involved is disease resistance. One way that higher plants repel the attacks of pathogenic micro-organisms is by the production of phytoalexins and B. J. Deverall presents here an up-to-date account of these fascinating plant substances. In the final chapter, W. G. H. Edwards turns our attention to chemical aspects of plant parasitism and refers to the terpenoid-like compounds which are excreted from the roots of certain crop plants only to trigger off the germination of *Orobanche* and other plant parasite seeds present in the vicinity.

The symposium programme was mainly organized by the Society officers but it is a pleasure to acknowledge also the assistance of Dr. E. M. Thain (Tropical Products Institute), Dr. R. T. Aplin (Oxford) and Professor J. B. Pridham (Royal Holloway College). The Society also wishes to thank the Unilever, I.C.I., Fisons, and Shell Corporations for generous financial assistance. The editor is particularly grateful to the contributors, who have made the task of editing this book such a pleasant one. He also acknowledges the valuable assistance of the Academic Press staff in preparing this book for publication.

December 1971   J. B. HARBORNE

Contents

CHAPTER 1

Some Observations on the Relationship Between Plants, Toxic Insects and Birds

Miriam Rothschild

CHAPTER 2

Chemical Aspects of Plant Attack by Leaf-cutting Ants

J. M. Cherrett

CHAPTER 3

Aphids as Phytochemists

H. F. van Emden

CHAPTER 4

Attractants and Repellents in Higher Animals

E. C. Bate-Smith

CHAPTER 5

The Chemistry of Flavour

T. A. Rohan

CHAPTER 6

Chemical Factors Affecting Selection of Food Plants by Ruminants

G. W. Arnold and J. L. Hill

CHAPTER 7

Cyanogenic Glycosides and Their Function

David A. Jones

CHAPTER 8

Aflatoxin and Related Mycotoxins

M. O. Moss

CHAPTER 9

Selenium Toxicity

Alex Shrift

CHAPTER 10

Toxic Amino Acids in the Leguminosae

E. A. Bell

CHAPTER 11
Toxicity and Metabolism of *Senecio* Alkaloids
A. R. Mattocks

I. Introduction 179
II. Chemical Structures 180
III. Toxic Effects of the Alkaloids. 182
IV. Metabolism of the Alkaloids 184
   A. Reasons for Believing that Toxicity is Due to Metabolites . . 184
   B. Possible Metabolic Pathways 185
V. Pyrrolic Derivatives as Toxic Metabolites. 188
   A. Reactivity as Alkylating Agents 188
   B. Pyrroles from Non-toxic Alkaloids 189
   C. Metabolic Formation of Pyrrole Derivatives 189
   D. Mechanism of Metabolic Pyrrole Formation 190
   E. Formation of Pyrrole Derivatives *in vivo* 191
VI. Toxicities of Semisynthetic and Synthetic Alkaloids and Pyrrole
   Derivatives 195
   A. Semisynthetic and Synthetic Pyrrole Derivatives . . . 195
   B. Synthetic Compounds Related to the Alkaloids Themselves . 197
VII. The Nature of the Toxic Reaction 197
VIII. Conclusions. 198
References 199

CHAPTER 12
Phytotoxins: An Ecological Phase of Phytochemistry
Cornelius H. Muller and Chang-Hung Chou

I. Introduction 201
II. Volatile Toxic Terpenes. 202
III. Water-borne Toxins 204
   A. California Chaparral 204
   B. Subhumid Deciduous Forest 210
IV. Role of Phytochemistry in Allelopathic Studies. . . . 213
   A. Implication of Specific Phytotoxins 213
   B. Ecology of Phytotoxins 214
Acknowledgements. 215
References 215

CHAPTER 13
Phytoalexins
B. J. Deverall

I. The Concept of Phytoalexins 217
II. Orchid Tubers, Cross Protection and Orchinol 219
III. Pisatin and Phaseollin as Phytoalexins 220
IV. A Range of Chemical Compounds as Phytoalexins . . . 222
V. Phytoalexins and Disease Resistance in Beans (*Vicia* and *Phaseolus*) . 225
   A. Broad Bean, *Botrytis*, Wyerone Acid and Wyerone . . 225
   B. Bean, *Colletotrichum* and Phaseollin 228
VI. Phytoalexins and Bacterial Diseases 229
VII. General Considerations 230
References 231

CHAPTER 14

Orobanche and Other Plant Parasite Factors

W. G. H. Edwards

CHAPTER 1

Some Observations on the Relationship Between Plants, Toxic Insects and Birds

MIRIAM ROTHSCHILD

Ashton, Peterborough, Northamptonshire, England

I. INTRODUCTION

The co-evolution of plants and insects has recently received considerable attention (Ehrlich and Raven, 1965; Brower, 1969; Fraenkel, 1969; Dethier *in* Sondheimer and Simeone, 1970; Macior, 1971) and special emphasis has been placed on the ability of certain aposematic (warningly coloured) butterflies to sequester and store secondary plant substances. The predators of Lepidoptera (at all stages of their life-cycle) have also been studied in greater detail than heretofore (Dempster, 1967, 1971).

In the last century when Bates (1862) first propounded the theory of mimicry it aroused considerable excitement because it was thought to demonstrate a situation in which natural selection could be seen in action. The sceptics argued against it on the grounds that birds did not prey on butterflies. Since then the pendulum has swung the other way and there is a tendency to discredit the theory of mimicry because it now appears that avian predators attack *all* butterflies, not only the innocuous mimics. Be that as it may, it seems generally agreed that birds are important predators of the Lepidoptera. Arthropods*

---

* Bacteria and virus infections are not considered here.

2

attack the early stages, i.e. the eggs and small caterpillars, but large caterpillars, the chrysalis and especially the winged imago are preyed on by birds.

II. THE ROLE OF BIRDS IN DETERMINING THE LIFE-STYLE OF WARNINGLY-COLOURED INSECTS

Entomologists, like birds, hunt by sight. Therefore the colours and patterns of aposematic insects have been studied more assiduously and with greater insight than their scents and sounds. Meldola (1882) nearly a hundred years ago suggested that it was the keen avian eye which selected the basic types of warning coloration characteristic of Danaid, Acraeid and Heliconid butterflies. We can add to this the aposematic patterns of diurnal moths such as Arctiids, Ctenuchids (= Syntomids) and Uraniids. Poulton (1932, 1937) agreed emphatically with Meldola on this point, and was convinced the theory explained the parallel evolution of the aposematic life-style of these unrelated groups. A comparison between insects feeding on tobacco and on various asclepiad plants provides an interesting example. Nicotine is, apparently, too general a poison to be sequestered and stored by insects. Two grasshoppers and several species of moths have been examined which feed on tobacco. In these cases nicotine is either rapidly excreted, or metabolized into a non-toxic derivative (Self *et al.*, 1964a,b). No host specific tobacco-feeding insect is aposematic,

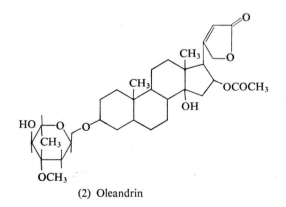

(1) Calotropin

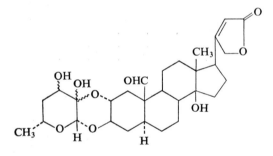

(2) Oleandrin

whereas a very high proportion of those feeding on the milk weeds (*Asclepias*) and on oleander (*Nerium*) are warningly-coloured. Moreover the majority of the latter species have the ability to sequester and store the cardiac glycosides present in the plants. These substances (e.g. I and II), which are generally bitter-tasting and toxic, render the insects distasteful and unacceptable to many predators (Marshall, 1902; Swynnerton, 1915, 1919, 1942; Jones, 1932; Fishelson, 1960; Brower *et al.*, 1968). Cardenolides have now been identified in five aposematic Lepidoptera, several Lygaeid bugs, Pyrgomorphid grasshoppers and beetles and one aphid. Any mutation facilitating recognition and avoidance by the avian predator will enormously enhance their survival value. There seems little doubt that it is the birds' superlative vision which has destined the tobacco-feeding grasshoppers, moths and beetles to be dull in colour and secretive in habit, and the majority of those insects feeding on *Asclepias* and oleander to be gay, eye-catching and self-advertising.

III. ADDITIONAL BENEFITS ACCRUING TO THE TOXIC PLANT FEEDERS

Trimen (1887), Poulton (1914) and Rothschild (1971) stressed the benefits accruing to the early stages of insects which have succeeded in establishing themselves on plants avoided by large herbivores. These animals, ranging from ungulates to voles, must destroy a vast number of eggs and caterpillars on vegetation which they consume. It has also been suggested that cardenolides possess bactericidal properties (Frings *et al.*, 1948) which can be passed on to the insects storing them. Once an insect is able to feed on such a plant, even if it fails to sequester and store the secondary plant substances (e.g. the Milkweed Beetle *Labidomera clavicollis* Kirby), various benefits accrue to it, some of which favour the development of the aposematic life-style. Thus, for the duration of the feeding period, the insect in question has a gut full of toxic and bitter-tasting material, and the presence of a complex of other aposematic insects, which renders its situation additionally favourable. Finally various warningly-coloured beetles, such as the seven-spot Ladybird (*Coccinella septempunctata* L.) and the Colorado Beetle (*Leptinotarsa decimlineata* Say) (Frazer and Rothschild, 1960; Hsiao and Fraenkel, 1969) possess independent repellent qualities, and they add to and enhance the assembly of aposematic insects feeding together in the area.

Obviously protection is never absolute, and the fact that gazelles (*Gazella gazella* Pallas) and the Hyrax* feed on oleander (personal observations), and certain birds (e.g. *Coturnix japonicus* Tem.) are insensitive to cardenolides (Kellett and Rothschild, 1971) and parasites destroy large numbers of aposematic insects, in no way invalidates Trimen's original suggestion that the

---

* The Hyrax (*Procavia*) eats other poisonous plants such as Euphorbiaceae, Solanaceae, the leaves of fig trees (*Ficus* sp.) (Mendelssohn, 1965) and *Phytolacca dodecandra* l'Herit (Sale, 1965).

initial success of the Danaids stemmed from their ability to lay and feed on protected plants. This was followed later by the refinement of plant poison
storage.

IV. Benefits Accruing to the Toxic Plant Host

At first sight it would seem that the relationship between plants and the
insect specialists which can browse on the toxic species, is a completely one-
sided affair—no benefits whatsoever accruing to the plant. But a closer look at
the situation may reveal a few hidden advantages. Many plants have evolved
specializations for attracting pollinators to their vicinity, one of which concerns morphological structures which simulate insects feeding on the flower
head. These range from the rather "crude" solitary purple flower in the centre
of the white inflorescence of certain Umbelliferae—nevertheless quite decep-
tively fly-like from a distance—to the extremely "sophisticated" model bee produced by the flowers of the genus *Ophrys*. The food plants of aposematic insects
need no such device, for, thanks to the sluggish habits of these insects, imposed
upon them by bird psychology and bird colour-vision, they sit quietly and
blatantly on flower heads, either on or near their food plant, for hours at a
time—thus effectively attracting other insects to the source of nectar and
pollen. Another benefit which undoubtedly accrues to to poisonous plants acting as hosts for aposematic insects is the reinforcement of their protective
odour. It is assumed that the powerful smell of *Asclepias* and *Aristolochia*
(Haase, 1896; Whittaker *in* Sondheimer and Simeone, 1970) acts as a warning
deterrent to large herbivores. Certainly grasshoppers of the genus *Poekilocerus*
smell strongly of their food plant (Beier, 1937; McCann, 1952; Fishelson,
1960; von Euw *et al.*, 1967) and so do various Danaid and Papilionid butterflies which feed on these plants. Ants, too, which associate with *Asclepias* and
thorn *Acacias*, possess powerful protective odours (Rothschild, 1964; Hocking,
1970), in certain cases strongly reminiscent of the food plant, which may be due
to the sequestering of essential oils from the plant in question, or provide an
example of the evolution of parallel scents (Rothschild, 1961). The former alternative seems probable in those cases where the defensive scent released by
aposematic insects seems to contain components from different toxic plants
with which they are associating at the time. Thus the seven spot ladybird prey-
ing on Aphids feeding on nettle (*Urtica dioica* L.), has a nettle-like scent component in its glandular secretion, but if feeding on *Aphis nerii* (Fonscolombe),
it smells strongly of oleander. An even more impressive example concerns the
grasshopper *Aularches miliaris* (L.). The nymphs of this species feed on *Hetero-
phragma* and smell strongly of that offensively odoriferous plant. As adults
they switch to feeding on *Gloriosa*, and at this period their smell changes and
comes to resemble that of the lily (McCann, 1952). Nymph and adult feed on
different plants and emit the smell of each in turn. In any case, the evocative
nature of these scents is very characteristic and must act as a powerful *aide*

mémoire to predators with prior experience of the prey in question. It has been claimed that these emanations can on occasion stimulate the sensation of *déjà vu* (Rothschild, 1967).

V. Secondary Plant Substances as Attractants and Repellents

Recently Fraenkel (1969) has evaluated the role of the secondary plant substances as attractants and repellents to herbivorous insects. He concludes that their importance in host selection is so overwhelming as to need no further proof. I would subscribe wholeheartedly to this generalization, but where the aposematic insects are concerned the situation is, as always, immensely complicated and varied. One may deduce that certain species of Danaids are actively attracted by plants which contain cardiac glycosides (Hopkins, 1927; Sevastopulo, 1971), for their alternative food plants are members of the Moraceae, such as *Streblus asper* Lour, which contain cardenolides and are extremely poisonous. On the other hand another genus of Danaids, *Amauris*, although they occasionally feed on the poisonous *Pergularia tomentosa* L. (and by inference can then store cardiac glycosides present in this plant, see Rothschild *et al.*, 1970b) usually oviposit and feed on those genera of Asclepiadaceae which do not contain these substances, for example *Tylophora*, *Cynanchum* and *Gymnema*. Perhaps *Amauris* is attracted first to the quite different secondary plant substances present in these genera, and only to the *Asclepias* containing cardenolides if the other plants are not available.

The situation with the Tiger Moths (Arctiids) is far more complicated. Many of these moths are largely polyphagous, but throughout the family there is a tendency to specialize on plants containing pyrrolizidine alkaloids.* Thus *Arctia caja* (L.) feeds on groundsel, and sequesters and stores the three *Senecio* alkaloids in this plant, but it does not oviposit or feed on *Senecio jacobaea* L. *Tyria jacobaeae* (L.) feeds on both *S. vulgaris* L. and *S. jacobaea*, and stores the alkaloids present in both plants. Different species of the genus *Utethesia* feed on *Crotalaria* and *Heliotropium*, plants from unrelated families which nevertheless contain pyrrolizidine alkaloids which these insects store. Various African Arctiids also feed on *Crotalaria* and store their alkaloids. However when the range of plants on which these moths are said to feed and specialize is examined, one is not only struck by the variety of poisonous species in the list, but also by the similarity between those selected and the food plants of aposematic insects from other Orders. Thus Lygaeid bugs, Pyrgomorphid grasshoppers as well as the Arctiid moths are recorded feeding on a wide range of *Asclepias* and Apocynaceae, and also *Senecio, Lantana, Euphorbia, Ricinus, Ipomoea, Gloriosa, Solanum, Datura*, species containing a wide variety of toxins. This suggests that poisonous plants may produce some

---

* The structures of many of the pyrrolizidine alkaloids mentioned in this chapter may be found in Chapter 11 (pp. 179–200).

generalized deterrent—a subtly different category of secondary plant substances—directed towards the larger herbivores—for example some scent or
odoriferous waxy exudate. This would be analogous to the generalized aposematic colour signals of toxic animals, which warn birds and other predators
of their noxious qualities. Insects may be able to take advantage of such a cue
and use it as a generalized *attractant*.

VI. Variation in the Occurrence and Pattern of Sequestered Plant Toxins

The study of the sequestering of secondary plant substances by aposematic
insects is in a very elementary stage, but it is already evident that the problem
is by no means simple. First and foremost the distribution of these substances
in plants is enormously varied in both time and space, and also, according to
the morphology, ecology and chemistry of the plant species, or population of
plants, or even the individual specimens concerned.

Thus, for example, the Milkweed, *Asclepias curassavica* L. apparently lacks
calactin in Brazil, but it is present in plants received from Israel, Trinidad and
the Royal Botanic Gardens, Kew (Santavy *et al.*, 1971 in preparation). The six
pyrrolizidine alkaloids present in ragwort (*Senecio jacobaea*) were found in
different proportions in plants collected at Weeting Heath, Suffolk, and
Ashton, Northants, a locality about one hundred miles further west
(Rothschild and Aplin, 1971a). The total alkaloid content was also greater in
the Weeting Heath plants. At Ashton itself the proportion of the alkaloids
varied in wild plants growing within 100 yards of each other. Moreover these
proportions changed significantly as the season advanced. For example,
seneciphylline was four times as high in August, as in June in ragwort, collected
from exactly the same site at Ashton. The three pyrrolizidine alkaloids in
groundsel (*Senecio vulgaris*) on the other hand varied little throughout the
year. Jones (1968, 1970 and this volume, p. 103) has shown that the
cyanogenic strains of clover and vetches occur at certain altitudes and not at
others, and such factors as temperature, age, or the absence and presence of
other plant species can influence or control the occurrence of the precursors of
HCN in such plants. Furthermore the location and distribution of secondary
plant substances in the tissues of plants is very imperfectly known. Thus in
certain species the roots or seeds or leaves or phloem may be rich in certain
cardenolides, and lacking in others. There are also specific differences in the
distribution of cardenolides in the same tissues.

These differences in the quantity and quality of the secondary plant substances are mirrored in certain insects feeding upon them (von Euw *et al.*,
1967; Brower *et al.*, 1968; Aplin *et al.*, 1968). Fed on lettuce, the grasshopper
Poekilocerus bufonius (Klug) contains only minute amounts of calactin and
calotropin (1) (probably carried over from the egg), and the caterpillar of the

Monarch butterfly (*Danaus plexippus* (L.)) fed on cabbage has none, or an insufficient quantity to cause emesis in bird predators.

We have noted that the storage sites of cardenolides in the plant host are not well known, and the same is true of the insects sequestering these substances. We also know little concerning their selectivity, although it is obvious species differ in this respect and also in their ability to concentrate various toxins. We found ten cardenolides in the body tissues of the Monarch butterfly (Reichstein *et al.*, 1968) reared on *A. curassavica*, two in *P. bufonius* (von Euw *et al.*, 1967), and two in *Aphis nerii* (Rothschild *et al.*, 1970a) reared on the same plant. The principal cardenolides in the first two species were calactin and calotropin, but the latter was missing in *A. nerii*.

Two Lygaeid bugs, *Caenocoris nerii* (Ger.) and *Spilostethus pandurus* L., reared on Oleander, both sequester and store heart poisons, but in the former, which is host specific, we found eleven cardenolides (over 30 were identified in the seeds of this plant by Jager *et al.*, 1959), but in the latter, which is polyphagous, only two. Neither insect apparently stored oleandrin (2)—the chief and most consistent cardenolide found in the oleander—but this is present in a Ctenuchid moth (Rothschild, 1972).

The Cinnabar moth caterpillar (*T. jacobaeae*) sequesters and stores all six pyrrolizidine alkaloids present in the food plant, but not in the same proportion as they occur in *S. jacobaea*. Thus 40–60% of these alkaloids present in the plant are jacobine, jacozine and jacoline, whereas these constitute but 10–15% of those found in the adult moth. On the other hand 60–70% of those sequestered and stored by the moth are senecionine, seneciphylline and integerrimine, which constitute 40–60% of those in the plant. A metabolite (20–25%) occurs in the moth, altogether lacking in the plant. The Arctiid *Amphicallia bellatrix* Dalm, which also sequesters and stores pyrrolizidine alkaloids, fed on *Crotalaria semperflorans* Vint., apparently only stores tricodesmine and crispatine, whereas the chief alkaloid of this host is said to be crosemperine. It is quite possible that an insect can transform one alkaloid from the plant host into another *in vivo*.

The eggs of the Tiger moth, *A. caja*, fed on groundsel contain *Senecio* alkaloids, but the eggs of the Cinnabar moth, *T. jacobaeae* do not. Some Arctiids are capable of sequestering and storing both alkaloids and cardiac glycosides (Rothschild and Aplin, 1971a,b). *Seriarctia echo* Kby stores the toxins from its cycad foodplant as the relatively harmless beta glycoside, leaving it to the host to reconvert it to the poisonous aglycone (Teas, 1967). The Common Blue butterfly (*Polyommatus icarus* (Rott.)), and a Weevil (*Hypera plantaginis* (Degeer)) which feed on cyanogenic strains of *Lotus corniculatus* L. metabolizes HCN to harmless thiocyanate by means of the enzyme rhodanese (Parsons and Rothschild, 1964). The same mechanism is employed by certain Hymenoptera and Tachinids which parasitize cyanogenic insects (Jones *et al.*, 1962), and by the larvae of the Botfly to combat HCN liberated when cyanogenic plants are digested by ruminants (Bertran, 1952).

The most efficient storage system so far investigated is that of the Pyrgomorphid grasshoppers already referred to, which can "carry over" small quantities of cardenolides—presumably via the egg—to the lettuce-fed F^1 and F^2 generations (von Euw et al., 1967).

VII. Variation in the Warning Signals of Aposematic Insects

It is obvious from the foregoing paragraphs that insects which store toxins derived directly from plants are bound to vary in the concentration of the poisonous substances in their body tissues. Thus the June population of Cinnabar caterpillars at Ashton contain less toxic material than the August population. The offspring of a female laying on groundsel (S. vulgaris) will lack three of the alkaloids present in those moths which feed up on ragwort (S. jacobaea) (Rothschild and Aplin, 1971a). The polyphagous Tiger moth A. caja, feeding on Foxglove (Digitalis) stores cardiac glycosides, and those feeding on groundsel, pyrrolizidine alkaloids (Rothschild and Aplin, 1971b). Coloration and scents and sounds also vary. The amount of white on the forewing of A. caja is a case in point, and so is the amount of red on the body of Parasemia plantaginis L: geographical and ecological differences occur in the concentration and quality of the repellent odour of the defensive gland secretion of Tiger moths (Rothschild, 1961, p. 103) and also in their ability to stridulate.

Reichstein showed that Gonolobus rostratus (Vahl), an alternative food plant for the Monarch butterfly (Danaus plexippus) in Trinidad, lacked cardenolides, and specimens reared on the plant proved palatable to birds (Brower, 1969). This butterfly, fed on different species of Asclepiad plants, displayed a palatability spectrum,* some individuals proving more palatable and more or less toxic than others. Variability of this type may play an important role in certain mimicry situations. Ford (1971) has drawn attention to the fact that in the absence of a model, or where it is very scarce, mimicry breaks down, and, for instance, the female morph of Papilio dardanus Brown, a classical Batesian mimic of Danaus chrysippus, is very variable in such circumstances. This is presumably due to relaxation of the intense selective pressures exercised by bird predators for "it is avian vision which checks variation within the gene complex" (Rothschild, 1971).

---

* The term "automimic" has been used to describe an insect specimen which, for example, contains fewer sequestered plant toxins in its tissues than another specimen (Brower, 1969, 1970; Brower et al., 1970) or which has temporarily depleted its store of poisonous secretions by discharging them at an aggressor (Eisner, 1970). It seems less confusing to restrict the term mimic to distinct species (polymorphic or otherwise) which resemble one or more other distinct species and not to use it (even with a prefix) to describe specimens of the same species which are variable. Similarly the term mimicry should be employed to describe interspecific not intraspecific relationships. It should be stressed that up to the present time only variations of toxicity within a population have been recorded, not true polymorphisms.

Owen (1970) in a most interesting paper, extends this concept to include a situation in which a common model is present, but of low toxicity. Thus he suggests that in West Africa the same model *D. chrysippus* may feed as larvae on plants which are less poisonous than in East Africa, and hence in this geographical area *Acraea encedon* L. (a polymorphic Müllerian mimic, an HCN secretor and unpalatable in its own right) does not develop a mimetic form resembling *D. chrysippus*. He attributes the relative dearth of Müllerian mimics associating with *D. chrysippus* in West Africa to the same cause. This is a fascinating idea and very plausible. It must be admitted, however, that it is by no means easy to visualize the role of the avian predator in this situation, for Swynnerton (1915, 1942) has shown that in Africa *Acraea* (he experimented with fourteen species, but not *A. encedon*) was as distasteful (sometimes more, and sometimes less so) than *D. chrysippus* itself, and also "caused nauseation". It may be true that the long awaited and long expected proof that insects can sequester and store toxic plant substances has greatly enhanced the theories of Wallace (1889), Trimen (1887) and Haase (1896), but it has certainly added an almost limitless series of possibilities and complications to the eternal triangle, plant → insect → bird, particularly in relation to warning coloration and mimicry.

VIII. SUMMARY

The bird predator has recently been neglected in the study of plant/insect co-evolution. It is suggested that avian vision imposes the aposematic life-style on insects which develop the ability to sequester and store harmful secondary plant substances, and accounts for the breakdown of mimicry in the absence of a model and, possibly, in localities where only feebly toxic models occur.

Warningly-coloured insects may bring some benefits to the plant host by the self-advertisement imposed upon them by predators, since by their behaviour they attract other pollinators to the site. They also reinforce protective plant odours, thus adding to the general aposematic effect of the complex.

The toxic substances within the plant host vary both in time and space, seasonally, ecologically, geographically, morphologically and chemically. Different species of insects store them selectively, or concentrate them in different proportions and in different body tissues, or alternately excrete, transform or metabolize all or some of them. The same species frequently mirrors the toxins present in different food plants; thus a polyphagous insect may store two or more unrelated poisons, such as pyrrolizidine alkaloids or cardiac glycosides, or lack plant toxins altogether.

Some unrelated plants may possibly develop a generalized warning system analogous to the aposematic colouring of distasteful animals, which serves as a deterrent to vertebrate herbivores and an attractant to certain herbivorous insects which can sequester and store their poisons.

2*

ACKNOWLEDGEMENTS

I am most grateful to Dr R. T. Aplin, Dr S. Duffey, Dr J. von Euw, Miss P. Fairweather, Mr D. S. Fletcher, Mr B. Ford, Professor T. Reichstein, Mr D. G. Sevastopulo, Dr J. A. Slater, Mr G. R. Cunningham van Someren and Professor G. C. Varley for much advice, information and help with collecting and rearing of material.

REFERENCES

Aplin, R. T., Benn, M. H. and Rothschild, M. (1968). Poisonous alkaloids in the body tissues of the cinnabar moth (*Callimorpha jacobaeae* L.). *Nature, Lond.* **219**, 747.

Bates, H. W. (1862). Contribution to an insect fauna of the Amazon Valley. Lepidoptera: Heliconidae. *Trans. Linn. Soc. Lond.* **23**, 495.

Beier, M. (1937). Eine neue Wehrdrüse bei saltatoren Orthopteren (Vorläufige Mitteilung). *C.R. int. Congr. Zool. Lisbon* (1935) **2**, 1458.

Bertran, E. C. (1952). Rhodanese in parasites. *An. Fac. Vet. Madrid* **4**, 334.

Brower, L. P. (1969). Ecological chemistry. *Scient. Am.* **220**, 22.

Brower, L. P. (1970). Plant poisons in a terrestial food chain and implications for mimicry theory. *In* "Biochemical Coevolution" (K. L. Chambers, ed.). Proc. 29 Ann. Biol. Colloquium 1968. Oregon State Univ. Press.

Brower, L. P., Ryerson, W. N., Coppinger, L. L. and Glazier, S. C. (1968). Ecological chemistry and the palatability spectrum. *Science, N.Y.* **161**, 1349.

Brower, L. P., Pough, F. H. and Meck, H. R. (1970). Theoretical investigation of automimicry. 1. Single trial learning. *Proc. natn. Acad. Sci. U.S.A.* **66**, 1059.

Dempster, J. P. (1967). The control of *Pieris rapae* with DDT. 1. The natural mortality of the young stages of *Pieris*. *J. appl. Ecol.* **4**, 485.

Dempster, J. P. (1971). Some effects of grazing on the population ecology of the cinnabar moth (*Callimorpha jacobaeae* L.). *Scientific Management of Animal and Plant Communities for conservation* (Symposium volume).

Dethier, V. G. (1970). Chemical interaction between plants and insects. *In* "Chemical Ecology" (E. Sondheimer and J. B. Simeone, eds.), Chap. 5. Academic Press, New York and London.

Ehrlich, P. R. and Raven, P. H. (1965). Butterflies and plants: a study in coevolution. *Evolution, Lancaster, Pa.* **18**, 586.

Eisner, T. (1970). Chemical defense against predation in arthropods. *In* "Chemical Ecology" (E. Sondheimer and J. B. Simeone, eds.), Chap. 8. Academic Press, New York and London.

Fishelson, L. (1960). The biology and behaviour of *Poekilocerus bufonius* Klug. with special reference to the repellent gland (Orth. Acrididae). *Eos, Madr.* **36**, 41.

Ford, E. B. (1971). "Ecological Genetics" (4th edition). Methuen, London.

Fraenkel, G. (1969). Evaluation of our thoughts on secondary plant substances. *Ent. exp. appl.* **12**, 473.

Frazer, J. F. D. and Rothschild, M. (1960). Defence mechanisms in warningly-coloured moths and other insects. *Proc. int. Congr. Ent. 11th Vienna* 1960, **3**, 249.

Frings, H., Goldberg, E. and Arentzen, J. C. (1948) Antibacterial action of the blood of the large milkweed bug. *Science, N.Y.* **108**, 689.

Haase, E. (1896). Mimicry in butterflies and moths. Researches on mimicry on the basis of a natural classification of the Papilionidae. Pt. 2. Researches on mimicry (translated by C. M. Child). Nagele, Stuttgart; Bailliere, Tindall & Cox, London.

Hocking, B. (1970). Insect associations with the swollen thorn acacias. *Trans. R. ent. Soc. Lond.* **122**, 211.

Hopkins, G. H. E. (1927). Insects of Samoa and other Samoan terrestrial arthropods. Part III. Lepidoptera. Fasc. 1 Butterflies of Samoa and some neighbouring island-groups. British Museum of Natural History, London, 1927–1935.

Hsiao, T. H. and Fraenkel, G. (1969). Properties of Leptinotarsin. A toxic haemolymph protein from the Colorado potato beetle. *Toxicon* **7**, 119.

Jager, H., Schindler, O. and Reichstein, T. (1959). Die Glykoside der Samen von *Nerium oleander*. L. *Helv. chim. Acta* **42**, 977.

Jones, D. A. (1968). On the polymorphism of cyanogenesis in *Lotus corniculatus* L. II. The Interaction with *Trifolium repens* L. *Heredity, Lond.* **23**, 435.

Jones, D. A. (1970). On the polymorphism of cyanogenesis in *Lotus corniculatus* L. III. Some aspects of selection. *Heredity, Lond.* **25**, 633.

Jones, D. A., Parsons, J. and Rothschild, M. (1962). Release of hydrocyanic acid from crushed tissues of all stages in the life-cycle of species of the Zygaeninae (Lepidoptera). *Nature, Lond.* **193**, 52.

Jones, F. M. (1932). Insect coloration and relative acceptability of insects to birds. *Trans. ent. Soc. Lond.* **80**, 345.

Kellett, D. N. and Rothschild, M. (1971). Notes on the reactions of various predators to insects storing heart poisons (cardiac glycosides) in their body tissues. *J. ent.* A (in press).

McCann, C. (1952). Aposematic insects and their food plants. *J. Bombay Nat. Hist. Soc.* **51**, 752.

Macior, L. W. (1971). Co-evolution of plants and animals. Systematic insights from plant-insect interactions. *Taxon* **20**, 17.

Marshall, G. A. K. (1902). Five years observations and experiments (1896–1901) on the bionomics of South African Insects, chiefly directed to the investigation of mimicry and warning colours. *Trans. ent. Soc. Lond.* **1902** (III), 287.

Meldola, R. (1882). Mimicry between butterflies of protected genera. *Ann. Mag. Nat. Hist. Ser.* **5**, **10**, 417.

Mendelssohn, H. (1965). Breeding the Syrian hyrax, *Procavia capensis syriaca* Schreber 1784. *Int. Zoo Yearbook* **5**, 116.

Owen, D. F. (1970). Mimetic polymorphism and the palatability spectrum. *Oikos* **21**, 333.

Parsons, J. A. and Rothschild, M. (1964). Rhodanese in the larva and pupa of the Common Blue Butterfly (*Polyommatus icarus* (Rott.)) (Lepidoptera) *Entomologists Gazette* **15**, 58.

Poulton, E. B. (1914). Mimicry in North American Butterflies: a reply. *Proc. Acad. nat. Sci. Philad.* (1914), 161.

Poulton, E. B. (1932). A hundred years of evolution. (Presidential address, Section D.) *Rep. Br. Ass.* **1931**, 71.

Poulton, E. B. (1937). The history of evolutionary thought as recorded in meetings of the British Association. (Presidential address). *Rep. Br. Ass.* **1937**, 1.

Reichstein, T., von Euw, J., Parsons, J. A. and Rothschild, M. (1968). Heart poisons in the Monarch butterfly. *Science, N.Y.* **161**, 861.

Rothschild, M. (1961). Defensive odours and Mullerian mimicry among insects. *Trans. R. ent. Soc. Lond.* **113**, 101.

Rothschild, M. (1964). A note on the evolution of defensive and repellent odours of insects. *Entomologist* **97**, 276.

Rothschild, M. (1967). Mimicry. *Nat. History* **76**, 44.

Rothschild, M. (1971). Speculations about mimicry with Henry Ford. *In* "Ecological Genetics and Evolution" (R. Creed, ed.). Blackwells, Oxford.

Rothschild, M. and Aplin, R. T. (1971a). Poisonous alkaloids in the body tissues of the Garden Tiger moth (*Arctia caja* L.) and the cinnabar moth (*Tyria* (= *Callimorpha*) *jacobaeae* L.) (Lepidoptera). *In* "Toxins of Plant and Animal Origin" (A. de Vries and E. Kochva, eds.). Gordon & Breach, London.

Rothschild, M. and Aplin, R. T. (1971b). Toxins in Tiger moths (Arctiidae: Lepidoptera). *In* "Pesticide Chemistry" (A. S. Tahori, ed.). Gordon & Breach, London.

Rothschild, M., von Euw, J. and Reichstein, T. (1970a). Cardiac glycosides in the Oleander aphid *Aphis nerii*. *J. Insect. Physiol.* **16**, 1141.

Rothschild, M., Reichstein, T., von Euw, J., Aplin, R. T. and Harman, R. R. M. (1970b). Toxic Lepidoptera. *Toxicon* **8**, 293.

Rothschild, M. (1972). Secondary plant substances and warning coloration in insects. *In* "Insect/Plant Relationships" (H. F. van Emden, ed.). *Symp. R. ent. Soc. Lond.* No. 6 (in press).

Sale, J. B. (1965). Hyrax feeding on a poisonous plant. *E. African Wildlife Journal* **3**, 127.

Santavy, F., von Euw, J. and Reichstein, T. (in preparation).

Self, L. S., Guthrie, F. E. and Hodgson, E. (1964a). Adaptation of tobacco hornworms to the ingestion of nicotine. *J. Insect. Physiol.* **10**, 907.

Self, L. S., Guthrie, F. E. and Hodgson, E. (1964b). Metabolism of nicotine by tobacco feeding insects. *Nature, Lond.* **204**, 300.

Sevastopulo, D. G. (1971). Food plants of the East African Lepidoptera (in preparation).

Swynnerton, C. F. M. (1915). Birds in relation to their prey: Experiments on Wood-Hoopoes, Small Hornbills and a Babbler. *Jl. S. Afr. Orn. Un.* **11**, 32.

Swynnerton, C. F. M. (1919). Experiments and observations bearing on the explanation of form and colouring, 1908–1913. *J. Linn. Soc.* (Zool.) **33**, 203.

Swynnerton, C. F. M. (1942). *In* G. H. D. Carpenter. Observations and experiments in Africa by the late C. F. M. Swynnerton on wild birds eating butterflies and the preference shown. *Proc. Linn. Soc. Lond.* **1941–1942**, 10.

Teas, H. J. (1967). Cycasin synthesis in *Seirarctia echo* (Lepidoptera) larvae fed methylazoxymethanol. *Biochem. biophys. Res. Commun.* **26**, 686.

Trimen, R. (1887). South African Butterflies: a monograph of the extra tropical species. Vols I–III. (London, 1887–1889.)

von Euw, J., Fishelson, L., Parsons, J. A., Reichstein, T., and Rothschild, M. (1967). Cardenolides (heart poisons) in a grasshopper feeding on milkweeds. *Nature, Lond.* **214**, 35.

Wallace, A. R. (1889). "Darwinism". Macmillan, London.

Whittaker, R. H. (1970). The biochemical ecology of higher plants. *In* "Chemical Ecology" (E. Sondheimer and J. B. Simeone, eds), Chap. 3. Academic Press, New York and London.

CHAPTER 2

Chemical Aspects of Plant Attack by Leaf-cutting Ants

J. M. CHERRETT

Department of Applied Zoology, University College of North Wales, Bangor, Wales

I. INTRODUCTION

In studies on the chemical relations between plants and the insects feeding upon them, a good deal of attention has been paid to the question of why some insects are polyphagous, feeding on many plants, whilst others are oligophagous or even monophagous, feeding on a restricted number, or even on one species only. House (1961) has stressed that the type of feeding relations adopted is a reflection of nutritional requirements, being metabolically advantageous to the species, and he quotes Gordon (1959) who emphasized that the ecologically significant question is not "what is an optimal diet for this animal?" but "what is the most deficient and unbalanced diet that the animal can tolerate without drastic reduction of its rate of growth and reproduction?" Dethier (1970) on the other hand, whilst accepting that some plants are toxic to insects or produce in them sub-optimal growth or development, considered nutritional requirements to be of secondary importance to the existence of "congruency" between the chemical composition of the plant and the chemosensory equipment of the insect. In his view, both are evolving by random mutation, and the feeding relations at any given time, as dictated by the pattern of phagostimulants and deterrents which are produced by the plant and perceived by the insect, is largely a matter of chance.

Jermy (1966) has suggested that the mechanism underlying increasing host specificity is an increasing sensitivity of the chemoreceptors to the feeding inhibitors produced by plants. This increasing awareness of the deterrents in plants generally is often associated with increasing sensory adaptation to specific phagostimulants produced by the small range of plant species on which the insect feeds.

Leaf-cutting, fungus-growing ants of the genera *Atta* and *Acromyrmex* have particularly interesting feeding relations which may throw some light on the problems referred to, because as Belt (1874) and Moeller (1893) demonstrated, the leaf material which they cut is used as a substrate for the culture of a fungus on which they feed. Weber (1966) has asserted that the ants eat only this fungus, although Echols (1966b) has demonstrated that *Atta texana* (Buckley) will drink soya bean oil, and Cherrett (1969) confirmed this for a series of oils with *Atta cephalotes* (L). Forel (1928) and Percy (1970) have reported that *A. cephalotes* readily drinks sucrose solutions, and the apparent drinking of sap oozing from cut stems is mentioned by Barrer and Cherrett (in press). It is not known if these dietary supplements are of importance in nature, and leaf-cutting ants are usually considered to be monophagous. However, Weber (1966) has successfully fed several species of the fungus *Lepiota* to laboratory ant colonies, and Craven *et al.* (1970) have recently described the existence of yeasts in the fungus gardens of *A. cephalotes* and *Acromyrmex octospinosus* (Reich), although their function is not clear. When the cutting of plant leaves for fungus substrate is considered, the animals are markedly polyphagous in the ecological sense.

This symbiotic relationship between leaf-cutting ants and their fungus has been studied in detail for the ant *Atta colombica tonsipes* Santschi (Martin and Carman, 1967). It appears that the fungus provides the ant with a diet of which more than 50% by dry weight is soluble nutrient, 27% of the dry weight being carbohydrate, 4·7% free amino acids, 13% protein and 0·2% lipid (Martin *et al.*, 1969). In particular, the fungus utilizes the cellulose in its medium, and Martin and Weber (1969) have estimated that at least 45% of the cellulose content of the leaf material placed on the fungus garden is consumed before the spent substrate is discarded. In return, the fungus is assiduously tended by the ants, and is transmitted from generation to generation by the queen, who carries fragments of mycelium in her infrabuccal pouch when on her mating flight (Huber 1905). The fungus is unknown outside ant nests (Weber, 1956). As the fungus is deficient in proteolytic enzymes, it does not grow well on material where protein is the main nitrogen source (Martin and Martin, 1970). The ants continually defecate on the fungus garden, and their faeces show marked protease activity (Martin and Martin, 1970) which compensates for the metabolic deficiencies of the fungus, so improving the competitive status of this fungus when exploiting the substrate of the fungus garden.

When cutting leaves as substrate for the fungus however, a very wide range of plant species is exploited (Moeller, 1893; De Souza, 1965; Weber, 1966), and in a nine week study of a colony of *A. cephalotes* in tropical rain forest, the present author found that 50% of the 72 identified woody plant species known to exist in the vicinity of the nest were cut for substrate (Cherrett, 1968). It is this catholicity of taste which makes leaf-cutting ants one of the most important general pests of agriculture, horticulture and forestry in the neotropics (Cramer, 1967; Cherrett and Sims, 1968). Despite the wide range of material exploited, leaf-cutting ants do exhibit preferences. *A. cephalotes* is largely restricted to broad leaved plants, and does not cut grasses; *A. laevigata* (Smith) will cut grasses and broad leaved plants; whilst *A. capiguara* Gonçalves is in the main restricted to grass (Amante, 1967). Within a given category of plants, some species appear immune to attack (Beebe, 1921; De Souza, 1965), whilst others, frequently introduced, cultivated species, appear particularly susceptible (Beebe, 1921; Belt, 1874; Forel, 1928).

The fact that leaf-cutting ants utilize for fungus substrate a very wide range of plant material but yet exhibit considerable selectivity, is the subject of this chapter.

II. ACCESSIBILITY OF CHEMICAL INFORMATION

Barrer and Cherrett (in press) have emphasized that there is a strong social component in the leaf-cutting activities of *A. cephalotes*. This is due not only to the scent trails which are laid by foragers to lead other workers to the cutting areas (Moser and Blum, 1963), but also to the fact that cuts in leaves are highly attractive to other ants, and induce further cutting in their vicinity. This may be in part a result of the mechanical advantage inherent in beginning a new cut from an existing cut, rather than from the leaf edge. However, when one millimetre of one edge of a leaf is cut away, so that no mechanical advantage exists, that edge is still much more attractive than the intact leaf edge. Clearly, additional factors must be involved, and Barrer and Cherrett (in press) suggest that sap oozing from the cut edge makes chemical information about the leaf more readily available to the ants. This implies that the intact leaf has some ability to "camouflage" its cutting-inducing chemicals!

III. CHEMICAL ARRESTANTS

General observations on leaf cutting behaviour suggest that the initial recognition of suitable material for cutting is through chemical arrestants (Beck, 1965), because filter paper discs treated with test chemicals were only examined and then picked up when the ants stumbled across them. No evidence of attraction at any appreciable distance has been observed. Cherrett and Seaforth (1970) conducted a preliminary analysis on grapefruit albedo (the white inner portion of grapefruit skins), and suggested that a variety of chemical groupings arrested *A. octospinosus*, whilst only two groupings arrested *A. cephalotes*. In both cases, the most active fraction consisted of a mixture of carbohydrates, amino acids and phenolic and other glycosides, which were not absorbed on a polyamide column. In the case of *A. cephalotes*, a synthetic mixture of sugars (fructose, glucose and sucrose) at approximately the physiological levels at which they exist in grapefruit albedo proved to be an effective arrestant, although this was not demonstrated for *A. octospinosus*. A comparable analysis of hibiscus (*Hibiscus rosa-sinensis*) flowers indicated that the same fraction was again the most active. Working with privet leaves, Barrer and Cherrett (in press) found that smearing sucrose solution onto the leaf surface greatly enhanced cutting by *A. cephalotes*.

IV. CHEMICAL REPELLENTS

It has frequently been reported that leaf-cutting ants preferentially select young leaves (Walter *et al.*, 1938; Fennah, 1950; Nelson, 1951), and although this may be in part due to their water content, or to mechanical considerations, changes in leaf chemistry appear to play an important role. Cherrett and Seaforth (1970) showed that discs cut from young grapefruit leaves were preferred by both *A. cephalotes* and *A. octospinosus* to discs cut from old leaves, and this also applied when a choice of alcoholic extracts of the two leaves was presented. In the case of *A. octospinosus*, there was further evidence that old leaves had developed repellents. Barrer and Cherrett (in press) found a similar preference for young privet leaves, and suggested that the old leaves may possess quantities of actively repellent chemicals (cutting inhibitors) which tend to counteract the influence of cutting stimuli.

Fresh grapefruit albedo is taken in preference to fresh flavedo (the yellow outer rind), although both are readily taken when the materials are dried. Cherrett (1969) suggested that this difference may be due to the presence of volatile repellents in the flavedo, and Cherrett and Seaforth (1970) produced some evidence that flavedo extracts were repellent to *A. octospinosus*, although this was not demonstrated for *A. cephalotes*.

In the present experiments, filter paper discs were dipped in a grapefruit albedo extract from which all lipid material had been removed. These discs were placed on glass microscope slides in pairs, and one disc was dipped in citrus oil, freshly obtained by scratching the flavedo of grapefruit with a pin, to puncture the oil sacs. These pairs of discs were placed in the foraging areas of laboratory colonies of *A. cephalotes*, and the disc first taken was noted. The glass slide was then removed, and a slide containing a new pair of discs substituted. In view of the volatility of some citrus oils, no slide was allowed to remain in the foraging arena for more than about 15 minutes before being renewed. From the results in Table I, it is clear that the oil from grapefruit flavedo is highly repellent, and this was confirmed by the behaviour of the ants which stopped short of the disc, held their antennae back, and made lunges at it. Indeed the two discs which were taken in replicate 3, were immediately dropped over the edge of the foraging table, no attempt being made to take them back to the nest as were the attractive control discs.

TABLE I

The effect of flavedo oils from grapefruit on the attractiveness to *Atta cephalotes* of filter paper discs impregnated with non-lipid albedo extract

	Number taken	
Replicate	Control discs	Discs plus flavedo oils
1	10	0
2	20	0
3	18	2
4	21	0
5	20	0

Chi square test assuming a 1:1 ratio.
Sum of $\chi^2 = (5$ d.f.$)$ 83·80 $(P < 0.001)$.
Pooled $\chi^2 = (1$ d.f.$)$ 83·18 $(P < 0·001)$.
Heterogeneity $\chi^2 = (4$ d.f.$)$ 0·62 $(P > 0·95)$.

V. THE BEHAVIOURAL RESPONSES OF THE ANTS

It seems likely that to leaf-cutting ants most natural materials contain a mixture of arrestant and repellent chemicals, and certainly their behavioural responses when selecting materials for fungus substrate appear to be complex. For laboratory colonies, marked fluctuations appear in the favourability of different substrate materials.

In a preliminary experiment, half a grapefruit skin, 6 young hibiscus leaves, and 4 hibiscus flowers with the sepals removed were presented to a colony of *A. octospinosus* at 09.00 hours on 40 days, spread over a period of 82 days. The positions of the 3 types of plant material were varied each day, and at 12.00 hours, the types of material being carried back by the first 30 ants observed were recorded. A high degree of variability in preference was apparent (see Table II), although there was an overall preference for hibiscus flowers. Because fresh material taken on different days and from several plants is likely to be variable in quality, a further experiment was performed on two sets of more homogeneous substrate.

In addition to cutting fresh leaves, flowers and fruit, leaf-cutting ants carry back to the nest, and utilize for fungus culture, a variety of dry materials such as grain (Wheeler, 1907), breakfast foods like cornflakes (Moser, 1967), cassava flour (Bates, 1884; Blanche, 1965) and dried citrus pulp (Echols, 1966a; Cherrett and Merrett, 1969). Similar materials were employed in this experiment, since a single batch of each could be used throughout. In the first trial, which was conducted on *A. cephalotes* for 30 consecutive days, the choice presented was "parboiled" rice grains, and two commercial breakfast foods—

TABLE II

Variability of preference by *Acromyrmex octospinosus* for three types
of fresh substrate[a]

	Grapefruit rind	Hibiscus flowers	Hibiscus leaves
Overall preference	387	612	201
Preferences on 3 selected dates			
14.11.66	2	28	0
11.1.67	24	3	3
13.1.67	9	5	16

$R \times C$ Chi square test $\chi^2 = 307 \cdot 8$ with 78 d.f. $P < 0 \cdot 001$.

[a] Number of fragments taken during the observation period.

TABLE III

Variability of preference by *Atta cephalotes* for three types of dried
substrate in two trials[a]

Trial 1	Porridge Oats	Bran Flakes	Parboiled Rice
Overall preference	5	146	2813
Preferences on 2 selected dates			
18.7.67	0	50	50
23.7.67	0	0	100

$R \times C$ Chi square test (columns one and two being pooled and two invalid rows omitted)

$$\chi^2 = 591 \cdot 3 \text{ with 27 d.f. } P < 0 \cdot 001$$

Trial 2	Dried Citrus pulp	"All Bran"	Parboiled Rice
Overall preference	2125	77	596
Preferences on 3 selected dates			
30.11.67	67	17	16
14.12.67	98	0	2
27.12.67	44	1	55

$R \times C$ Chi square test (omitting column two, and one invalid row)

$$\chi^2 = 386 \cdot 2 \text{ with 28 d.f. } P < 0 \cdot 001$$

[a] Number of fragments taken during the observation period (see text).

bran flakes, and porridge oats. In the second trial, the choice was "parboiled" rice, the commercial breakfast food "all bran" and dried citrus pulp. In both cases the number of ants taking each material was recorded for 100 animals, or if there was little activity, for the final hour of the daily two hour observation

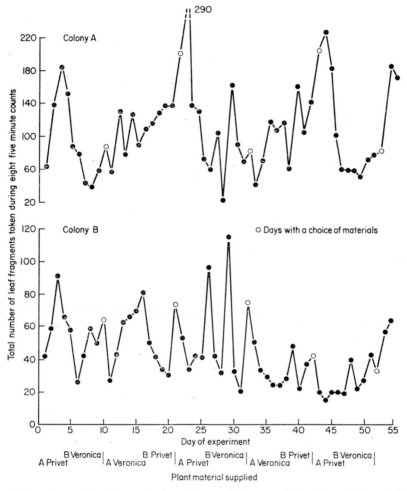

FIG. 1. Foraging activity of two colonies of *Atta cephalotes* supplied with two different plant materials.

period. In Table III, a similar result to that of Table II was obtained, with overall preferences appearing despite pronounced preference variability from day to day.

This variability in preference for different types of substrate has also been demonstrated for different chemical fractions from a single type of substrate, Cherrett and Seaforth (1970) describing variability in the preferences shown for single batches of chemical extracts. This was noted from day to day, for a given nest of both *A. cephalotes*, and *A. octospinosus*; it was noted between different colonies of the same species, and it was also observed to occur between the 2 species. They attributed these differences in response to social phenomena evolved by the ants.

In an attempt to investigate further this phenomenon of preference heterogeneity, two similar sized laboratory colonies of *A. cephalotes* were subjected to the following feeding regimes. Colony A was provided daily with an excess supply of fresh privet leaves (*Ligustrum ovalifolium* Hasskarl), all old leaves being removed before the fresh supply was added, whilst at the same time, Colony B was being given a daily excess of fresh veronica leaves (a variety of *Veronica angustifolia*). On each day, 8 counts were made at 15-minute intervals, of the number of leaf fragments passing into each nest, the time of beginning these counts being standardized to within 2 hours. After 9 or 10 days of this treatment, each nest was given a choice between veronica leaves and privet leaves, the total number of leaves given being the same as usual. On this occasion, the numbers of each type of leaf which were taken were recorded. For the next 10 days, the leaf species were reversed, a nest which had previously been given privet, being provided only with veronica, and *vice versa*. The whole experiment was run for 55 consecutive days, and the type of substrate given to each colony was changed on 5 occasions.

The amount of foraging undertaken by each colony varied greatly from day to day for reasons unknown, but there is some evidence that the amount of foraging recorded on the "changeover" days, when a choice of two substrates

TABLE IV

The preferences of two laboratory colonies of *Atta cephalotes* on dates
when they were given a choice of two leaf species[a]

Date	Colony A			Colony B			Between Colony heterogeneity	
	Previous leaves provided	Pr	Ve	Previous leaves provided	Pr	Ve	χ^2	P
8.11.70	Pr	11	80	Ve	32	32	25·08	<0·001
19.11.70	Ve	123	84	Pr	24	50	14·85	<0·001
30.11.70	Pr	25	65	Ve	54	21	30·31	<0·001
10.12.70	Ve	134	81	Pr	6	37	31·86	<0·001
20.12.70	Pr	13	82	Ve	27	7	47·54	<0·001

Between date heterogeneity for each colony

$$\chi^2 \quad 132\cdot03 \quad 57\cdot72$$
$$P \quad <0.001 \quad <0.001$$

[a] Total number of leaf fragments taken during eight 5-minute counts spread over 2 hours.
Pr = Privet; Ve = Veronica.

was available, was greater than that on the preceding day (Fig. 1). On such days, foraging increased on 9 occasions, and decreased on only 1. A Chi Square Test showed that this ratio was significantly different from $1:1$ ($\chi^2 = 4.9$ with 1 d.f. $P < 0.05$).

The preferences of each colony for either privet or veronica leaves on the five "changeover" days (Table IV) show a very high degree of heterogeneity. In 9 out of 10 cases, the ants chose preferentially the leaves which they had not been given on the previous 10 days, and a Chi Square test showed this to be significantly different from a $1:1$ ratio ($\chi^2 = 4.9$ with 1 d.f. $P < 0.05$). As the leaves which had been given to the two colonies were different, this meant that on any one "changeover" date, the two colonies were exhibiting opposed preferences. This colony heterogeneity was also highly significant.

VI. DISCUSSION

Although no evidence of an olfactory attractant (Hsiao, 1969) causing the leaf cutting ants *A. cephalotes* and *A. octospinosus* to orientate towards leaf material has been observed, it appears that chemical arrestants (Beck, 1965) are important in initiating the sequence of behaviour leading to leaf cutting. As these animals collect leaf material to act as a fungus substrate, but feed in the main on the fungus which they grow, the sequence of stimuli necessary for complete feeding activity is probably more complex than is usual in phytophagous insects (Munakata, 1970). Thus, the stimuli required to initiate leaf cutting, the carrying of the material back to the nest, and its incorporation into the fungus garden, must be distinct from those which enable the ant to recognize the fungus, begin feeding on it, and maintain that feeding.

Despite this fact that vegetable material is only cut as a fungus substrate, some plant defence mechanisms still appear to operate. The ability of intact leaves of privet partly to conceal or "camouflage" their cutting-inducing chemicals is a simple mechanism, which has been noted previously (Ghent, 1960; Sudah, 1969). It is interesting to note that in both of the leaf species studied (privet and grapefruit), old leaves were much less readily cut than were young, a phenomenon which seems in part to be the result of increased chemical repellency. Feeny (1967) has shown that old oak leaves have a deleterious effect on the growth of larvae of the winter moth (*Operophtera brumata*), and it would be interesting to know how general is the phenomenon of older leaves developing increasing chemical resistance to defoliating insects. Repellents in grapefruit flavedo present a more specialized case, as the peel oil tested was obtained from special oil sacs, where the oil appears to be maintained under some pressure. Kefford and Chandler (1970) state that 93–95% of peel oil consists of limonene, and Cherrett (1969) showed that this chemical was not attractive to *A. cephalotes*, although it was not tested for repellency. The albedo is of course highly attractive to the ants (Cherrett and Seaforth, 1970). As grapefruit is exotic to the areas where leaf cutting ants occur, the development of pressurized oil sacs in the flavedo must be regarded as a general defence mechanism, which happens to be effective against these species.

If leaf-cutting ants use leaf material simply as a substrate for fungus culture, it is not clear if there is a functional basis behind their avoidance of certain plant materials. It may be that the ants find repellent any substrates which are toxic to the fungus, or on which the fungus does not grow well, or it may be that the ants simply find some chemicals to be so repugnant that they cannot cut material containing them, even though it might be suitable as fungus substrate.

The variability in preferences shown by leaf-cutting ants for both natural materials and chemical fractions, may be explicable in part by the tendency to get tired of material taken day after day, and so select novel alternatives when they become available. Tinbergen (1960) has described a similar phenomenon in great tits, as have Rodgers and Rozin (1966) for rats fed on a diet deficient in thiamine. In arthropods, the reverse has been more generally described, Turnbull (1960) demonstrating that a field population of the spider *Linyphia triangularis* (Clerck) was highly conservative in the prey species which it accepted, being initially reluctant to accept different species as the seasons progressed, whilst Jermy *et al.* (1968) were able to implant certain feeding preferences in larvae of *Manduca sexta* (Johan) and *Heliothis zea* (Boddie) simply by feeding them on that plant for a short initial period.

In the case of *A. cephalotes*, this apparent preference for novelty, as Dethier (1970) has described this type of reaction, will tend to maintain the remarkable degree of "polyphagy" in the collection of substrate upon which many authors have commented. The function of such behaviour is not clear. It may be that the fungus performs best on a mixed substrate, and detecting deficiencies in a single substance, the ants seek new material, in the way described by Rodgers and Rozin (1966) for rats. In culture, however, laboratory colonies have survived well on only one or two exotic species of plant leaves. The maintenance of a high degree of "polyphagy" does carry important ecological implications, as it will help to prevent the over exploitation of preferred plant species in the vicinity of these long-lived nests, and so spread the grazing pressure more evenly over the available resources.

A preference for novelty by leaf-cutting ants does make it difficult to study the phytochemical attractants and repellents to which they respond, and as their feeding behaviour involves the nutrition of another organism between the collecting of raw materials by the ants, and their ingestion of food, their feeding relations become difficult to interpret in terms of the current general theories about monophagy and polyphagy set out in the introduction.

ACKNOWLEDGEMENTS

The author is grateful to Mr D. Jones for technical assistance, and to Dr R. G. Gibbs, Mr J. Hobart and Dr D. J. Peregrine for reading the manuscript and for valuable discussion. The studies were supported by a grant from the Overseas Development Administration of the Foreign and Commonwealth Office.

REFERENCES

Amante, E. (1967). *Biologico* **33**, 113–120.
Barrer, P. M. and Cherrett, J. M. (In press).
Bates, H. W. (1884). "The Naturalist on the River Amazons", 5th edition. John Murray, London.
Beck, S. D. (1965). *A. Rev. Ent.* **10**, 207–232.
Beebe, W. (1921). "Edge of the Jungle." Duell, Sloan and Pearce, New York.
Belt, T. (1874). "The Naturalist in Nicaragua". Bumpus, London.
Blanche, D. (1965). Congrès de la protection des cultures tropicales. *C. r. trav., Marseille* 449–454.
Cherrett, J. M. (1968). *J. Anim. Ecol.* **37**, 387–403.
Cherrett, J. M. (1969). *Trop. Agric., Trin.* **46**, 81–90.
Cherrett, J. M. and Merrett, M. R. (1969). *Trop. Agric., Trin.* **46**, 221–231.
Cherrett, J. M. and Seaforth, C. E. (1970). *Bull. ent. Res.* **59**, 615–625.
Cherrett, J. M. and Sims, B. G. (1968). *J. agric. Soc. Trin.* **68**, 313–322.
Cramer, H. H. (1967). "Plant Protection and World Crop Production". 'Bayer' Pflanzenschutz, Leverkusen.
Craven, S. E., Dix, M. W. and Michaels, G. E. (1970). *Science, N.Y.* **169**, 184–186.
De Souza, L. F. (1965). Divulgaçào agronómica (Shell). **14**, 23–29.
Dethier, V. G. (1970). *In* "Chemical Ecology" (E. Sondheimer and J. B. Simeone, eds), pp. 83–102. Academic Press, New York and London.
Echols, H. W. (1966a). *J. econ. Ent.* **59**, 628–631.
Echols, H. W. (1966b). *J. econ. Ent.* **59**, 1336–1338.
Feeny, P. P. (1968). *J. Insect Physiol.* **14**, 805–817.
Fennah, R. G. (1950). *Proc. agric. Soc. Trin.* **50**, 312–326.
Forel, A. (1928). "The Social World of the Ants", Vol. 2 (translated by C. K. Ogden). Putnam's Sons, London.
Ghent, A. W. (1960). *Behaviour* **16**, 110–148.
Gordon, H. T. (1959). *Ann. N.Y. Acad. Sci.* **77**, 290–351.
House, H. L. (1961). *A. Rev. Ent.* **6**, 13–26.
Hsiao, T. H. (1969). *Entomologia exp. appl.* **12**, 777–788.
Huber, J. (1905). *Biol. Zbl.* **25**, 606–619 and 625–635.
Jermy, T. (1966). *Entomologia exp. appl.* **9**, 1–12.
Jermy, T., Hanson, F. E. and Dethier, V. G. (1968). *Entomologia exp. appl.* **11**, 211–230.
Kefford, J. F. and Chandler, B. V. (1970). "The Chemical Constituents of Citrus Fruits". Academic Press, New York and London.
Martin, J. S. and Martin, M. M. (1970). *J. Insect Physiol.* **16**, 227–232.
Martin, M. M. and Carman, R. M. (1967). *Science, N.Y.* **158**, 531.
Martin, M. M. and Weber, N. A. (1969). *Ann. Ent. Soc. Am.* **62**, 1386.
Martin, M. M. and Martin, J. S. (1970). *J. Insect Physiol.* **16**, 109–119.
Martin, M. M., Carman, R. M. and MacConnell, J. G. (1969). *Ann. ent. Soc. Am.* **62**, 11–13.
Moeller, A. (1893). *Bot. Mitt. Trop.* **6**, 1–127.
Moser, J. C. (1967). *Nat. Hist., N.Y.* **76**, 32–35.
Moser, J. C. and Blum, M. S. (1963). *Science, N.Y.* **140**, 1228.
Munakata, K. (1970). *In* "Control of Insect Behaviour by Natural Products" (D. L. Wood *et al.* eds). Academic Press, New York and London.

Nelson, H. S. (1951). The San Tome Experimental Station Bull. No. 1. Mene Grande Oil Co.
Percy, H. C. (1970). Unpublished M.Sc. Thesis, University of Wales.
Rodgers, W. and Rozin, P. (1966). *J. comp. physiol. Psychol.* **61**, 1–4.
Sudah, I. M. (1969). Unpublished M.Sc. Thesis, University of Wales.
Tinbergen, L. (1960). *Archs neerl. Zool.* **13**, 265–343.
Turnbull, A. L. (1960). *Can. J. Zool.* **38**, 859–873.
Walter, E. V., Seaton, L. and Mathewson, A. A. (1938). United States Department of Agriculture, Circular 494.
Weber, N. A. (1956). *Proc. Xth Int. Congr. Ent.* **2**, 439–473.
Weber, N. A. (1966). *Science, N.Y.* **153**, 587–604.
Wheeler, W. M. (1907). *Bull. Am. Mus. nat. Hist.* **23**, 669–807.

CHAPTER 3

Aphids As Phytochemists

H. F. VAN EMDEN

*University of Reading, Horticultural Research Laboratories,
Shinfield Grange, Shinfield, Berkshire, England*

I. The Phenomenon of Host Specificity

"God also said . . . All green plants I give for food to the wild animals, to all birds of heaven, and to all reptiles on earth, and every living creature" (Genesis I: 30). It is striking that living creatures have in fact a rather restricted food range yet, at the same time, there seem to be very few flowering plants which are not susceptible to at least one plant-feeding insect. Thorsteinson (1960) wrote: "These considerations clearly imply the existence of mechanisms that participate in the effective allocation of grazing privileges to herbivorous insects. The outcome of this allocation yields the patterns of 'host selection' or 'food-plant preferences' that we observe among phytophagous insects." Lipke and Fraenkel (1956) regarded the subject of host selection by insects as "the very heart of agricultural entomology".

As aphids are probably the insect group of greatest agricultural importance and at the same time clearly demonstrate "allocation of grazing privileges" it seems appropriate for this symposium to discuss what is known about the phytochemical characteristics which influence infestation of plants by aphids.

The close association of different aphid species with individual plant species is apparent in the fact that the common names of aphids seem more botanically orientated than is probably true for any other group of insects. Thus we refer to "willow aphid", "sycamore aphid", "spruce aphid", "pea aphid" and "cherry blackfly" to name but a few.

Even among six economically important aphid species affecting six important crops, a wide variation in host specificity is apparent (Table I, p. 33).
There is a range from e.g. the cabbage aphid (*Brevicoryne brassicae*) which is
virtually restricted to crucifers and is monoecious (no alternation between a

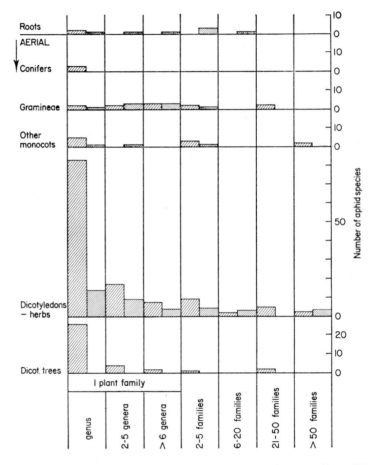

Fig. 1. Host distribution of aphid species among various vegetation types (data of Kennedy
et al., 1962). Shaded blocks, monoecious aphids; dotted blocks, dioecious aphids
(primary *or* secondary hosts are scored)

winter woody and a summer herbaceous host) to the shallot aphid (*Myzus
ascalonicus*) which is monoecious yet rather polyphagous and to the lettuce root
aphid (*Pemphigus bursarius*) which has a very restricted host range yet is hetero-
ecious (alternating between poplar and lettuce). From data by Kennedy *et al.*
(1962) it is possible to summarize graphically the host plant distribution of a
large proportion of the British aphid fauna (Fig. 1). Not only are the majority of

these aphids monoecious, but also most are restricted to one genus of plants. It is unusual to find aphid species which bridge plant genera, even within one botanical family.

II. HOST SELECTION BEHAVIOUR

Such distribution of aphids among plant species (Fig. 1) occurs despite the fact that the arrival of alate aphids in an area of vegetation is undirected. With most aphids, the production of alate progeny on both winter and summer hosts is induced by overcrowding stimuli (van Emden *et al.*, 1969) following a rapid succession of parthenogenetic generations of apterous viviparous females. Typically the newly moulted adult alatae fly from the plant on which they were reared if the weather in the morning following their moult is suitable; they are then soon carried upwards to considerable heights (e.g. 1500 m) by atmospheric turbulence (Taylor, 1965). By the time turbulence ceases in the evening and the aphids descend, they have been carried a considerable horizontal distance. They thus descend onto host and non-host plants and have been found to alight on both with equal frequency (Kennedy *et al.*, 1959a,b; Müller, 1958, 1962), although there is some evidence that an aphid can direct its landing in the last few centimetres of descent under the influence of visual stimuli (Moericke, 1955a,b, 1957).

Having landed, the aphid "probes" by repeated short (up to a few minutes) insertions of its stylets at different sites (Hennig, 1963, 1966). Such probing behaviour has been observed in many aphid species, and presumably supplies the aphid with information about the internal chemical and physical properties of the substrate. This information then determines whether the aphid takes off again and leaves the plant or whether it remains. Thus the plant species distribution of aphids is far more due to differential re-take-off than to differential alighting.

The extensive work of Müller on the black bean aphid (*Aphis fabae*) provides an example of this general pattern (Müller, 1962) (Fig. 2). Both bean and beet are hosts for this species, whereas mustard is a non-host. Alightment on all three species was approximately equal, with a ratio close to 50:50% in each paired comparison. Mustard plants were rejected very quickly after only a few probes, and the proportion of alighted aphids remaining after two minutes was very low on this plant species. In fact, alighted aphids continue to leave all plants over a period of many hours and, even on host plants, only a small proportion remain after twenty-four hours. Thus many aphids fail to find a suitable site even where such sites may be supposed to be abundant. Müller (1958) showed that the 1:3·5 difference in initial infestation by *A. fabae* between the resistant bean variety Rastatt and the susceptible variety Schlanstedt was related to re-take-off rates as high as 99 and 90% respectively.

Although some groups of aphids (e.g. the Adelgidae) feed predominantly in parenchymatous tissue, it is generally supposed that, particularly in the

Aphididae, most species feed on phloem sap (Auclair, 1963). This has been challenged for some species, particularly for the peach-potato aphid (*Myzas persicae*) (Lowe, 1967); however, the following discussion of the probing process nevertheless assumes phloem feeding as normal.

The route of aphid stylets to the phloem may be intercellular or intracellular, probably more often the former. Navigation of the stylets is achieved by mov-

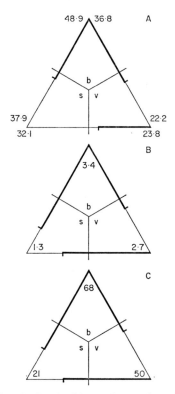

Fig. 2. Alighting and re-take-off of *Aphis fabae* on host and non-host plants (data of Müller, 1962), presented as proportions of unity (the thin line intersecting each side of the triangle indicates the 50% point). A, number of aphids alighting; B, mean number of probes before re-take-off; C, alighted aphids remaining after two minutes; b, *Beta* (sugar-beet); v, *Vicia* (bean); s, *Sinapis* (mustard); thick lines, proportional response

ing the curved tips against each other (Miles, 1958), so that they slide round at a "T-junction" of cell walls in the appropriate direction. That the stylets travel through the mesophyll towards the phloem by itself suggests that the aphid can monitor the chemistry of the substrate and detect chemical gradients, although there is rather little experimental evidence favouring this idea. In fact both Ehrhardt (1961) and Hennig (1963) (Fig. 3) were unable to detect uptake of ^{32}P from plants by *Megoura viciae* and *A. fabae* until the phloem was reached. McLean and Kinsey (1968), however, claimed to detect uptake (by recording induced voltage variations associated with liquid flow through the stylets) by the pea aphid (*Acyrthosiphon pisum*) from both epidermal and mesophyll tissues (Fig. 4). It takes the aphid at least fifteen minutes (and often longer) to reach the phloem (Hennig, 1966; van Hoof, 1958), and there is a correlation

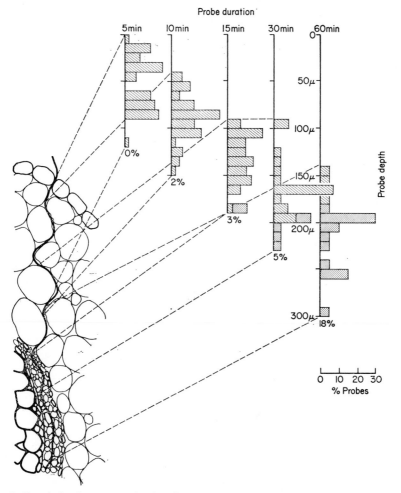

Fig. 3. Correlation between probe duration and depth (data of Hennig, 1966). %, per cent probes where ^{32}P was ingested; dotted columns, probes which reached the phloem

between probe time and probe depth (Fig. 3). The various stages of stylet penetration (plant surface, penetration of epidermis, movement of the stylets through the mesophyll, ingestion from the phloem) present the aphid with a series of "take it or leave it" situations; thus whether the aphid stays or departs may be determined without a probe, after a short probe (as little as 30 seconds to 1 minute) or after the phloem is reached.

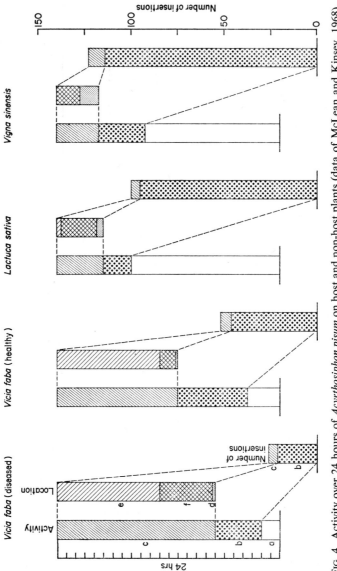

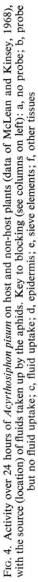

FIG. 4. Activity over 24 hours of *Acyrthosiphon pisum* on host and non-host plants (data of McLean and Kinsey, 1968), with the source (location) of fluids taken up by the aphids. Key to blocking (see columns on left): a, no probe; b, probe but no fluid uptake; c, fluid uptake; d, epidermis; e, sieve elements; f, other tissues

Although aphids probably reject plants following difficulty in reaching the phloem (e.g. McMurtry and Stanford, 1960), chemical characteristics of the plant substrate are of particular importance. Rejection of a substrate on chemical grounds can be exceedingly rapid; Marek (1961) obtained rejection of acidified filter papers by *Myzus ascalonicus* within ten seconds of a probe, whereas sucrose solution on filter paper was accepted. The tip of the labium may have chemoreceptors and certainly the mandibles are enervated internally with a pair of dendrites (Forbes, 1966). This enervation may enable navigation of the stylet tips, but may alternatively enable the aphid to detect chemical properties (e.g. acidity) of the substrate at the stylet tips; however, Miles (1968) points out that a sense of taste would seem inadequately served by two nerve processes alone. Miles also points out that, if fluids are drawn up the food canal to contact chemoreceptors in the roof of the pharynx and are then discharged again, the conflicting evidence from difference methods of detecting food uptake by aphids during probing (see earlier) is resolved. Fluid would move up the food canal and produce a variation in induced voltage, yet no radioactive tracer would be ingested by the aphid.

McLean and Kinsey (1968) used their electronic induced voltage recording method to examine probing behaviour of *Acyrthosiphon pisum* caged on plants of varying host quality for twenty-four hours. This gives a striking picture (Fig. 4) of the relation of probing activity and plant host status, for the aphids were forced to stay on plants from which they might otherwise have departed after only a very few probes. On unsuitable plants (e.g. *Vigna sinensis*) the aphids spent over half the time walking and changing position between an enormous number of short probes. Fluid was taken up during less than 10% of the probes, and then almost entirely from the epidermis and other non-phloem tissue. At the other extreme, diseased *Vicia faba* appeared the most acceptable host. Almost 22 hours were spent with the stylets inserted, and the small number of probes (only about one-tenth of that on *V. sinensis*) reflects the greater length of each individual probe. Over two-thirds of the time was occupied by a few insertions when fluid was also taken up. Most of this uptake was from the phloem and least from the epidermis. *Lactuca sativa* and healthy *Vicia faba* occupied intermediate positions on these criteria of host status.

Thus an overall picture of probing behaviour in relation to host status of the plant emerges. On unsuitable hosts, the aphid is restless, and leaves after a few short probes in rapid succession. On a suitable substrate the probes are longer and result in greater fluid uptake, particularly from the phloem.

Although some stylet withdrawals are responses to physical characteristics of the substrate, such characteristics are probably of no importance at least once the phloem has been reached. The phloem may be reached even on non-host plants (Ehrhardt, 1961) and therefore before acceptance of the plant has been determined. Thus chemical characteristics of the plant are crucial at this stage, and probably also during penetration of the mesophyll if one bears in mind that uptake of fluid appears to occur at this stage.

III. The Chemical Criteria of Host Selection

There has been much controversy, although this is now mainly in the past, over the chemical criteria involved in host selection by plant feeding insects. The early discovery by Verschaeffelt (1910) that mustard oil glycosides influence host selection and larval feeding of the cabbage white butterfly (*Pieris rapae*) drew attention to what are often called "secondary substances of no nutritive value". Such compounds are often characteristic of botanical groupings and include particular glycosides, alkaloids, flavonoids and terpenes. Fraenkel (1959) argued that many secondary substances which now encourage feeding by particular pests are in fact evolved by plants to resist insect attack; pest races adapted to such compounds now accept them as recognition stimuli. This argument receives wide acceptance. Verschaeffelt's (1910) classic discovery led several workers (particularly Fraenkel, 1956) to claim in addition that secondary substances were predominant in determining host selection. This claim was partly based on the association of certain insects with plants well-known to contain characteristic compounds such as amygdalin (some Rosaceae), salicin (Salicaceae) and methylonylketones (Umbelliferae). In support of this claim, Fraenkel (1953) collected data of analyses of green leaves which suggested that all the major nutrients would be present in all leaves in adequate amounts for any plant-feeding insect. However, Kennedy and Booth (1951) had been impressed by the variations in plant susceptibility to aphids that could occur within one plant species and that could be correlated with physiological changes in the plant (particularly changes associated with leaf growth and senescence). They had proposed a "dual discrimination theory" of aphid host selection: the sensory stimuli for aphid feeding responses were not solely secondary substances, but included universal substances of fundamental nutritional importance to plants and insects alike or at least token stimuli physiologically associated with such universal substances.

Since the early 1950's, some experimental evidence has accumulated concerning the relative roles of secondary substances and major nutrients in determining susceptibility of plants to aphids. Much of this evidence expresses plant susceptibility to aphids in terms other than host selection (e.g. adult longevity, aphid multiplication rate), but it seems likely (van Emden *et al.*, 1969) that selection and subsequent multiplication (at least on summer hosts) are favoured by the same plant characteristics.

The remainder of this paper is limited to observations on only two aphid species which both colonize Cruciferae. There are four main reasons for this limited choice:

(1) Because of the early work of Verschaeffelt (1910) identifying the importance of mustard oil glycosides for cabbage caterpillars, the attention of aphid workers concerned with secondary substances has been focused on brassicas.

(2) Sinigrin (1) is commercially available in pure form for *in vitro* experiments on gustatory stimuli.

TABLE I

Some examples of host plant distribution of pest aphid species

Aphid	Winter hosts			Summer hosts					
	Euonymus	Populus	Prunus	Solanum	Brassica	Beta	Chrysanthemum	Lactuca	Triticum
Aphis fabae	+					+			
Brevicoryne brassicae					+				
Myzus ascalonicus				+	+	+	+	+	
Myzus persicae			+	+	+	+	+		
Pemphigus bursarius		+						+	
Sitobion avenae									+

3

$$CH_2{=}CH{-}CH_2{-}N{=}C\diagup^{S-Glc}_{\diagdown OSO_2OK}$$

(1) Sinigrin

$$CH_2{=}CH{-}CH_2{-}N{=}C{=}S$$

(2) Allylisothiocyanate

(3) The use of mustard as a condiment has resulted in the availability of analytical techniques for determining total allylisothiocyanate (2) concentrations in plant material.

(4) The two aphid species concerned seem an ideally contrasting pair for such a study. The cabbage aphid (*Brevicoryne brassicae*) is monoecious and virtually restricted to the Cruciferae. It appears to select its host plants on the basis of their taxonomic affinities and (van Emden, 1966) is relatively little affected by variations in the physiological condition of the host plant. By contrast, the peach potato aphid (*Myzus persicae*) has several dioecious clones overwintering on some *Prunus* spp., and its summer populations develop on a wide range of plants including several unrelated and important crops such as potato, sugar beet, brassicas and chrysanthemums (Table I). The aphid is much affected by the physiological condition of the plant substrate.

It is tempting to suppose that "taxonomic host selection" involves secondary substances alone, and that "physiological host selection" is entirely dependent on nutrients. However, this ignores the fact that "taxonomic" and "physiological" plant variations are not necessarily different in kind but are, as far as the probing aphid is concerned, artificial classifications of mechanical and phytochemical variation. Indeed, it is apparent from recent experiments (see below) that no such distinction can be made and it is not safe to claim that variations within one plant species in susceptibility to insects cannot involve secondary substances.

IV. DIRECT EVIDENCE FROM ARTIFICIAL DIETS

The development of chemically defined ("artificial") diets for aphids (Mittler and Dadd, 1962; Auclair and Cartier, 1963) has made it possible to vary the concentration of single compounds in aphid food and to test the effect of this variation on aphids. Thus many problems of whole plant experiments (especially internal correlation between compounds) are avoided, though, even with artificial diets, one cannot increase the concentration of one compound without adding less of something else or altering the balance between the constituents.

The interaction of the main nutrients (sucrose and amino acids) with sinigrin has not, in fact, been studied by any one worker. However, a picture of this interaction can be obtained (Fig. 5) by combining the results of several workers (Wensler, 1962; Mittler and Dadd, 1964; Moon, 1967; Wearing, 1968; van Emden, unpublished data from a student class experiment). With artificial diets, van Emden measured settling of aphids in a four-choice chamber whereas

Mittler and Dadd, also Moon, used single choice tests; Wearing recorded aphid longevity. Wensler measured the settling of aphids on bean leaves, some of which had taken up dissolved sinigrin via the petioles. However, parallel results were obtained by measuring probe number and duration on damp pith discs covered with a thin colloidin membrane, and therefore her results for bean (containing natural sucrose and amino acids) are included here. Figure 5a presents the possible combinations of choices in the form of the interaction of sucrose and amino acids with (back wall) and without (front wall) added sinigrin. Possibly because of preconceived ideas, the various workers have offered rather different choices to the two aphid species. Thus there is little data on amino acids and sinigrin respectively for *B. brassicae* and *M. persicae*. The fragmentary results available do, however, suggest a marked difference in

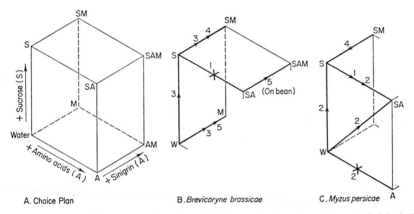

A. Choice Plan   B. *Brevicoryne brassicae*   C. *Myzus persicae*

FIG. 5. Responses of *Brevicoryne brassicae* and *Myzus persicae* to nutrients and sinigrin in artificial diet. Data taken from: 1, van Emden (unpublished); 2, Mittler and Dadd (1964); 3, Moon (1967); 4, Wearing (1968); 5, Wensler (1962). Thick lines, comparisons made; arrows point to more favourable combination; S, sucrose; A, amino acids; M, sinigrin

the response of the two aphids. *B. brassicae* (Fig. 5b) responds positively to sinigrin and probably also to sucrose, but there is no evidence of a selection for amino acids. *M. persicae* (Fig. 5c) responds positively to sucrose, but also to amino acids in the presence of sucrose. If anything, there appears to be a negative response to sinigrin. Thus the most favourable combination for *B. brassicae* appears to be *sucrose + sinigrin* (with or without amino acids) and for *M. persicae sucrose + amino acids − sinigrin*.

Mittler and Dadd (1965) studied the duration of probes by *M. persicae* on various artificial diets with particular reference to the response to sucrose and amino acids. Their results agree with those obtained later by McLean and Kinsey (1968) with their electronic recording method on plants (see earlier)— i.e., that length of probe increases with suitability of the substrate. Acid solutions or hard objects (e.g. test tube walls) were rejected far more rapidly than even a dry membrane. On water or a mixture of six amino acids there were a few probes longer than those on a dry membrane. These longer probes (more than four minutes) were the most frequent insertions on 20 % sucrose, and were increased still further by the addition of amino acids. Such long probes were even more frequent again on the "complex" diet, which included vitamins and various salts.

V. EVIDENCE FROM WHOLE PLANTS ON THE PERFORMANCE OF APHIDS AFTER SETTLING

Little evidence has been obtained from whole plants on the *relative* role of nutrients and secondary substances in determining plant susceptibility to aphids. Following the initial work of Lindemann (1948) and Mittler (1958), a great deal of data has accumulated to link aphid susceptibility of plants within a single species to high soluble nitrogen levels (reviewed by van Emden *et al.*, 1969), it being generally though only theoretically assumed that carbo-hydrate levels in plants are unlikely to be limiting. In whole plants, vitamins have been almost entirely neglected by aphidologists, presumably because of difficulties of quantitative estimation.

van Emden and Bashford (1969) endeavoured to create variations in both the soluble nitrogen and secondary substance concentrations in brassica plants by using fertilizer regimes and studying different leaf ages (by position) on the plant. Total allylisothiocyanate was found to decrease (nearly halve) down the plant, the largest decrease occurring between "young" (276 ppm) and "medium" (167 ppm) leaves. The fecundities of both *M. persicae* and *B. brassicae* were recorded and, at any one leaf age, were correlated with soluble nitrogen concentration. Both aphids, however, showed a greater fecundity per unit soluble nitrogen the older the leaf (Fig. 6); this led the authors to suggest that any comparison of leaf "ages" was limited by the deficiencies of whole leaf analyses where phloem feeding insects were concerned. However, the in-crease in aphid fecundity per unit soluble nitrogen on older leaves was so marked for *M. persicae* (actually higher fecundities were recorded on old than young leaves in spite of the low soluble nitrogen concentrations in old leaves) that it was suggested that high mustard oil glycoside concentrations might have made the otherwise suitable young leaves unfavourable for this aphid.

At the same time, it was clear that total soluble nitrogen was rather a rough measure of soluble nitrogen, and the amino acids and some amides were there-fore studied in more detail (van Emden and Bashford, 1971) in an experiment with plant age as the "treatment" variable. The "performance" of *B. brassicae* and *M. persicae* was assessed over 4 days by the aphids' mean relative growth rate (RGR), which can be regarded as a constant for each individual up to the moult to adult (van Emden, 1969). The plants were then harvested and the amino acids of the soluble fraction in mature leaves (on which the aphids had been caged) were analysed. Asparagine and glutamine could not be separated; the complex of these two compound is designated "amide" in this paper. Each plant in the experiment thus provided data for RGR of the two aphids as well as the concentration of the individual amino acids. The data were subjected to principle component analysis, the results of which can be summarized briefly as follows:

(1) A very high degree of multiple correlation could be obtained between RGR and only three amino acids for each aphid.

(2) Both aphids correlated positively with "amide" concentration; this may explain the general correlation found to exist between susceptibility of plants to aphids and total soluble nitrogen.

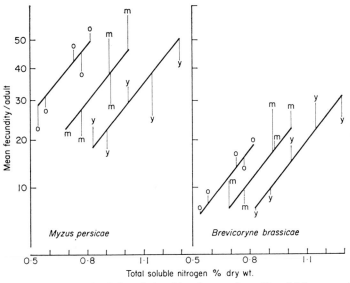

FIG. 6. Simple linear models of the relationship of mean fecundity of *Myzus persicae* and *Brevicoryne brassicae* to soluble nitrogen levels and leaf age in the Brussels sprout plant (data of van Emden and Bashford, 1969). y, m, o, points observed experimentally on young, medium and old leaves

(3) Each aphid species correlated positively with further amino acids, but also negatively with some amino acids including γ-aminobutyric acid (3) in the case of *M. persicae*.

$$NH_2—CH_2—CH_2—CH_2—CO_2H$$
(3) γ-Aminobutyric acid

(4) The greater "nutrient" sensitivity of *M. persicae* than *B. brassicae* was again apparent.

From the results of several similar experiments involving different "treatments", it is possible to begin extending the tentative regression equations for RGR of *M. persicae* and *B. brassicae* proposed by van Emden and Bashford (1971). Figure 7 illustrates, with reference to two chrysanthemum varieties, how the "amino acid prediction" for RGR of *M. persicae* is arrived at. The work is, however, still at an early stage and further modifications will almost certainly be made as more information accrues on upper and lower thresholds for aphid response to individual amino acids. The concentration of each "favourable" amino acid is multiplied by the appropriate regression coefficient (left hand column for each variety) to give the "credit" side of the plant's contribution to aphid RGR. The "debit" side (the "unfavourable" amino acids) is subtracted in the right hand column; the residue (shaded) is taken as the "predicted"

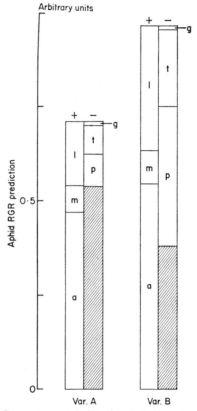

FIG. 7. Prediction of *Myzus persicae* RGR (shaded blocks) on two chrysanthemum varieties (A more susceptible than B), showing the contribution of: a, amide; g, γ-amino-butyric acid; l, leucine; m, methionine; t, tyrosine; p, proline

RGR. One feature apparent from Fig. 7 is how much the unfavourable amino acids contribute to the relative resistance of the variety on the right.

As the aphids appear to be sensitive to the amino acid spectrum of the plant as well as to secondary substances (cf. work on artificial diets described earlier), it was decided to combine these two variables in a single experiment involving a replicated trial of twenty-four plants covering various aspects of the phyto-chemistry of plant taxonomy and physiology. The results of this experiment are reported here for the first time. Four plant species in different genera were sown (two Cruciferae, *Brassica rapa* (turnip) and *Sisymbrium officinale*; one Compositae, *Lapsana communis*; one Rosaceae, *Poterium polygamum*). Each plant species was presented to the aphids in six physiologically different guises (three plant ages, each grown in either JI1 compost or "hedgerow" soil). The RGR of both *B. brassicae* and *M. persicae* was then measured on each plant (Fig. 8).

The distinction between the aphid species in terms of response to plant "taxonomy" and "physiology" was apparent even before the plants were analysed. *B. brassicae* showed a high taxonomic selectivity and did not survive on the non-cruciferous plants. In spite of the physiological variations built into the experimental plants, there was relatively little variation in the RGR of the aphid.

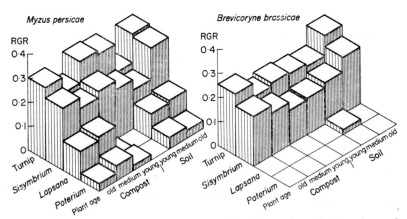

FIG. 8. RGR of *Myzus persicae* and *Brevicoryne brassicae* on four plant species, each of various physiological conditions

By contrast, *M. persicae* succeeded in colonizing most of the "treatments" of the experiment. Moreover, there was a high variation in RGR of the aphid, considerable differences occurring among the six physiological variants of each plant species.

Chemical analysis of the plants was limited to the two crucifer species, as the secondary substances of *Lapsana* and *Poterium* could not be analysed. Mature leaves from all "treatments" of the crucifers were analysed quantitatively for amino acids and total allylisothiocyanate. The amino acid quantities were then used in the regression equations developed from previous experiments to obtain an "amino acid prediction" for RGR for each aphid on each plant "treatment".

On separate graphs (Fig. 9), the performance (mean RGR) of each aphid species could then be plotted against the "amino acid prediction" for RGR and total allylisothiocyanate. In the graphs, the thick black lines include the area covered by the experimental points. There was extensive overlap between the points of the two plant species on both the amino acid and allylisothiocyanate scales, indicating that both types of compound varied with "physiology" as well as plant species. The experimental points were used for partial regression, and the two-dimensional slope of the response of RGR to the two plant variables was calculated.

In Fig. 9, the slopes have been extrapolated beyond the experimental points and shaded to illustrate more clearly the generalized response of the aphids to the two types of plant constituent. The following points can be made from the results of this experiment:

(1) Variation in both secondary substances and nutrients is important in determining the performance of both aphid species.

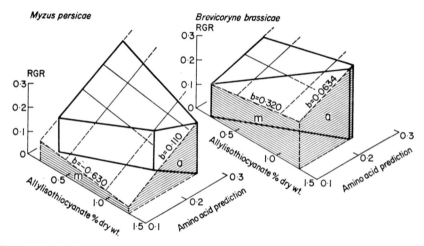

FIG. 9. Simple linear regression model of the relationship between RGR of *Myzus persicae* and *Brevicoryne brassicae* to amino acids (a) and total allylisothiocyanate (m) in crucifer plants. Thick lines, area covered by the experimental data of Fig. 8

(2) Both types of variation are involved in either "taxonomic" or "physiological" plant comparisons.

(3) Both aphids respond positively to improvements in the amino acid characteristics of the substrate. Such "improvement" involves changes in both "favourable" and "unfavourable" amino acids.

(4) The characterization of an "improved" amino acid substrate seems similar (at least for *M. persicae*) in different plant species (Brussels sprout, turnip, *Sisymbrium* and chrysanthemum).

(5) *M. persicae*, which is fairly polyphagous, is more strongly influenced than *B. brassicae* by changes in amino acids and over a greater concentration range.

(6) *M. persicae* responds negatively to an increase in allylisothiocyanate, in contrast with the host (crucifer) restricted *B. brassicae* which shows a positive response.

(7) For both aphids, the two chemical constituents (amino acids and allyliso-thiocyanate) of the plant which were analysed accounted for a large proportion of the observed variability in mean performance of the aphids. This proportion was 81 and 75% of the treatment sum of squares for *M. persicae* and *B. brassicae* respectively.

VI. CONCLUSIONS

The results of experiments on whole plants confirm the hypothesis based on artificial diet work that both nutrients and secondary substances are involved in determining susceptibility of plants to aphids. *B. brassicae* appears influenced by amino acids though it may not select for these. It can be provisionally suggested (based on the regression equations for this species in the whole plant experiments reported here and which equations included threshold values for most amino acids) that the amino acids which correlate with the performance of *B. brassicae* fluctuate little and occur at "favourable" concentrations in most crucifer plants within wide limits of physiological condition.

The amino acids "important" for *M. persicae* fluctuate greatly with condition of the host plant and the aphid appears to select for such compounds particularly by selecting plants or sites on plants of characteristic tissue age. Such selection on crucifers also appears to involve secondary substances, and the young leaves may be avoided because of their high allylisothiocyanate content.

ACKNOWLEDGEMENT

The author's work described under Section V, "Evidence from whole plants" has been generously financed by the Nuffield Foundation.

REFERENCES

Auclair, J. L. (1963). Aphid feeding and nutrition. *A. Rev. Ent.* **8**, 439–490.
Auclair, J. L. and Cartier, J. J. (1963). Pea aphid: rearing on a chemically defined diet. *Science, N.Y.* **142**, 1068–1069.
Ehrhardt, P. (1961). Zur Nahrungsaufnahme von *Megoura viciae* Buckt., einer phloemsaugende Aphide. *Experientia* **17**, 461–463.
Forbes, A. R. (1966). Electron microscope evidence for nerves in the mandibular stylets of the green peach aphid. *Nature, Lond.* **212**, 726.
Fraenkel, G. S. (1953). The nutritional value of green plants for insects. *Trans. 9th Int. Congr. Ent., Amsterdam, 1951* **2**, 81–88.
Fraenkel, G. S. (1956). Insects and plant biochemistry: the specificity of food plants for insects. *Proc. 14th Int. Congr. Zool., Copenhagen, 1953*, 383–387.
Fraenkel, G. S. (1959). The raison d'être of secondary plant substances. *Science, N.Y.* **129**, 1466–1470.
Hennig, E. (1963). Zum Probieren oder sogenannten Probesaugen der schwarzen Bohnenlaus (*Aphis fabae* Scop.). *Entomologia exp. appl.* **6**, 326–336.
3*

Hennig, E. (1966). Zur Histologie und Funktion von Einstichen der schwarzen Bohnenlaus (*Aphis fabae* Scop.) in *Vicia faba*-Pflanzen. *J. Insect Physiol.* **12**, 65–76.

Kennedy, J. S. and Booth, C. O. (1951). Host alternation in *Aphis fabae* Scop. I. Feeding preferences and fecundity in relation to the age and kind of leaves. *Ann. appl. Biol.* **38**, 25–64.

Kennedy, J. S., Booth, C. O. and Kershaw, W. J. S. (1959a). Host finding by aphids in the field. I. Gynoparae of *Myzus persicae* (Sulzer). *Ann. appl. Biol.* **47**, 410–423.

Kennedy, J. S., Booth, C. O. and Kershaw, W. J. S. (1959b). Host finding by aphids in the field. II. *Aphis fabae* Scop. (Gynoparae) and *Brevicoryne brassicae* L.; with a re-appraisal of the role of host-finding behaviour in virus spread. *Ann. appl. Biol.* **47**, 424–444.

Kennedy, J. S., Day, M. F. and Eastop, V. F. (1962). "A conspectus of aphids as vectors of plant viruses." Commonw. Inst. Ent., London, 114 pp.

Lindemann, C. (1948). Beitrag zur Ernährungsphysiologie der Blattläuse. *Z. vergl. Physiol.* **32**, 112–133.

Lipke, H. and Fraenkel, G. S. (1956). Insect nutrition. *A. Rev. Ent.* **1**, 17–24.

Lowe, H. J. B. (1967). Interspecific differences in the biology of aphids (Homoptera: Aphididae) on leaves of *Vicia faba*. I. Feeding behaviour. *Entomologia exp. appl.* **10**, 347–357.

McLean, D. L. and Kinsey, M. G. (1968). Probing behaviour of the pea aphid, *Acyrthosiphon pisum*. II. Comparisons of salivation and ingestion in host and non-host leaves. *Ann. ent. Soc. Am.* **61**, 730–739.

McMurtry, J. A. and Stanford, E. H. (1960). Observations of feeding habits of the spotted alfalfa aphid on resistant and susceptible alfalfa. *J. econ. Ent.* **53**, 714–717.

Marek, J. (1961). Über das Einstich- und Saugverhalten der Zwiebellaus, *Myzus ascalonicus* Doncaster. *Z. PflKrankh. PflPath. PflSchutz.* **68**, 155–165.

Miles, P. W. (1958). The stylet movements of a plant-sucking bug, *Oncopeltus fasciatus* (Dall.) (Heteroptera: Lygaeidae). *Proc. R. ent. Soc. Lond.* (A) **33**, 15–20.

Miles, P. W. (1968). Insect secretions in plants. *A. Rev. Phytopath.* **6**, 137–164.

Mittler, T. E. (1958). Studies on the feeding and nutrition of *Tuberolachnus salignus* (Gmelin) (Homoptera, Aphididae). II. The nitrogen and sugar composition of ingested phloem sap and excreted honeydew. *J. exp. Biol.* **35**, 74–84.

Mittler, T. E. and Dadd, R. H. (1962). Artificial feeding and rearing of the aphid *Myzus persicae* (Sulzer), on a completely defined synthetic diet. *Nature, Lond.* **195**, 404.

Mittler, T. E. and Dadd, R. H. (1964). Gustatory discrimination between liquids by the aphid *Myzus persicae* (Sulzer). *Entomologia exp. appl.* **7**, 315–328.

Mittler, T. E. and Dadd, R. H. (1965). Differences in the probing responses of *Myzus persicae* (Sulzer) elicited by different feeding solutions behind a parafilm membrane. *Entomologia exp. appl.* **8**, 107–122.

Moericke, V. (1955a). Über die Lebensgewohnheiten der geflügelten Blattläuse unter besonderer Berücksichtigung des Verhaltens beim Landen. *Z. angew. Ent.* **37**, 29–91.

Moericke, V. (1955b). Über das Verhalten phytophager Insekten während des Befallsflugs unter dem Einfluss von weissen Flächen. *Z. PflKrankh. PflPath. PflSchutz* **62**, 588–593.

Moericke, V. (1957). Der Flug von Insekten über pflanzenfreien und pflanzenbewachsenen Flächen. *Z. PflKrankh. PflPath. PflSchutz* **64**, 507–514.

Moon, M. S. (1967). Phagostimulation of a monophagous aphid. *Oikos*, **18**, 96–101.

Müller, H. J. (1958). The behaviour of *Aphis fabae* in selecting its host plants, especially different varieties of *Vicia faba*. *Entomologia exp. appl.* **1**, 66–72.

Müller, H. J. (1962). Über die Ursachen der unterschiedlichen Resistenz von *Vicia faba* L. gegenüber der Bohnenblattlaus, *Aphis* (*Doralis*) *fabae* Scop. VIII. Das Verhalten geflügelter Bohnenläuse nach der Landung auf Wirten und Nichtwirten. *Entomologia exp. appl.* **5**, 189–210.

Taylor, L. R. (1965). Flight behaviour and aphid migration. *Proc. N. cent. Branch ent. Soc. Am.* **20**, 9–19.

Thorsteinson, A. J. (1960). Host selection in phytophagous insects. *A. Rev. Ent.* **5**, 193–218.

van Emden, H. F. (1966). Studies on the relations of insect and host plant III. A comparison of the reproduction of *Brevicoryne brassicae* and *Myzus persicae* (Hemiptera: Aphididae) on Brussels sprout plants supplied with different rates of nitrogen and potassium. *Entomologia exp. appl.* **9**, 444–460.

van Emden, H. F. (1969). Plant resistance to *Myzus persicae* induced by a plant regulator and measured by aphid relative growth rate. *Entomologia exp. appl.* **12**, 125–131.

van Emden, H. F. and Bashford, M. A. (1969). A comparison of the reproduction of *Brevicoryne brassicae* and *Myzus persicae* in relation to soluble nitrogen concentration and leaf age (leaf position) in the Brussels sprout plant. *Entomologia exp. appl.* **12**, 351–364.

van Emden, H. F. and Bashford, M. A. (1971). The performance of *Brevicoryne brassicae* and *Myzus persicae* in relation to plant age and leaf amino acids. *Entomologia exp. appl.* **14**, 349–360.

van Emden, H. F., Eastop, V. F., Hughes, R. D. and Way, M. J. (1969). The ecology of *Myzus persicae*. *A. Rev. Ent.* **14**, 197–270.

van Hoof, H. A. (1958). Onderzoekingen over de biologische overdracht van een non-persistent virus. *Meded. Inst. PflZiekt. Onderzoek, Wageningen*, no. 161, 96 pp.

Verschaeffelt, E. (1910). The cause determining the selection of food in some herbivorous insects. *Proc. Acad. Sci. Amsterdam* **13**, 536–542.

Wearing, C. H. (1968). Responses of aphids to pressure applied to liquid diet behind parafilm membrane. Longevity and larviposition of *Myzus persicae* (Sulz.) and *Brevicoryne brassicae* (L.) (Homoptera: Aphididae) feeding on sucrose and sinigrin solutions. *N. Z. Jl Sci. Technol.* **11**, 105–121.

Wensler, R. J. D. (1962). Mode of host selection by an aphid. *Nature, Lond.* **195**, 830–831.

CHAPTER 4

Attractants and Repellents in Higher Animals

E. C. BATE-SMITH

*ARC Institute of Animal Physiology, Babraham,
Cambridge, England*

I. GENERAL INTRODUCTION

In the case of animals other than insects, the specific interactions between plants and animals are almost entirely concerned with the latter eating the former. As will be apparent in this and the following two chapters, it is generally agreed that "flavour" provides the basis for the selection of plant species as food by animal species. As Rohan mentions in Chapter 5, flavour is a complex of taste and aroma. Gustation and olfaction are indeed chemical senses. As Arnold and Hill express it in Chapter 6 of this volume, "Animals live in worlds dominated by chemical communication systems."

Much of the subject matter in this and the following chapters is therefore devoted to the triangular relationship between palatability, chemistry and species. Arnold and Hill deal with the study of palatability by the methods of the animal behaviourist, Rohan deals with the relationship between palatability and chemistry—the criterion of palatability being the human subject. The present author endeavours to show how the human subject is indispensable as an interpreter of the behaviour of the animal in terms of the palatability and chemistry of the plant.

The use of the sense organs of the human subject as reliable—in fact the *only* really valid—instruments in the analysis of flavour has only just become acceptable to, and established as a routine in, the food industries. At last the nose is being applied to the effluent of the gas chromatograph. What is up to now totally lacking, a codex of relationship between chemical constitution and odour properties, must surely be forthcoming from these activities. Rohan mentions in Chapter 5 of this volume the essay in the description of odour by Harper *et al.* (1968). This attempts to trace the evolution of odour qualities in organic molecules by simple *stepwise* changes in substitution from the inodorous methane and water molecules. This seems still to be the most promising approach and would fulfil Rohan's plea for less ambitious attempts than those aiming to arrive at universal relationships between flavour and chemical structure. It would have the additional advantage, to those so minded, of being admirably adapted to computerization.

This represents a kind of "morphology" of odour characters based on chemical structures. It implies a multiplicity of odour taxa just as susceptible to systematic arrangement in terms of odour characters as are plant species in terms of structural characters. But to achieve the construction of such a system would require a body of odour observers trained in odour recognition as assiduous and dedicated as the amateur and professional plant taxonomists; and no doubt equally unlikely to agree on their systems when they have got them.

II. On Eating Plants

A. FREEDOM OF CHOICE

It has to be recognized that, so far as chemical aspects of higher animal-plant interactions are concerned, there is extraordinarily little known. However, there is quite a lot that can be inferred, from what is known about the chemistry of the plants that animals eat, as to the nature of the attractant and repellent constituents present in them.

To make a start with the human animal, and specifically our present selves, one is immediately struck with the very narrow range of plants from which we select our staple foods; the very few even of those which are not staple that are at all frequently eaten. Some of the reasons for our restricted menu emerge when man's eating habits are traced back in history. This exercise has recently been undertaken by the Institute of Archaeology in London in a book edited by Ucko and Dimbleby (1969) entitled "The Domestication and Exploitation of Plants and Animals."

TABLE I

Human		Other animal	
Primitive	Free	Wild	Free
	Restricted		Restricted
Sophisticated	Civilized	Domesticated	Restricted
	Urbanized		Confined

The progressive restriction in choice of plant foods both of man and animals can usefully be discussed under the headings set out in Table I.

There are probably no longer any living representatives of the primitive, free human animal, but there may still be pockets of humans living in the primitive conditions of Neolithic man in Africa, South-East Asia, and elsewhere. The Bushmen, for instance "only make use of a (culturally) selected number of edible berries within the terrain which they occupy" (Ucko and Dimbleby, 1969). The Veddhas of Ceylon are mainly meat-eating; "wild yams, truffles, kurrakhan, maize, fruits and certain jungle edible flowers make up their menu" (Williams, 1963). (Note the inclusion of the exotic cereals, kurrakhan and maize, in their presently available plant species!) This brief list must, however, be very attenuated when one takes into account the wide range of plants that find their way into the markets of neighbouring cultures—so many, in fact, that even trained botanists are often not immediately able to recognize them (R. L. M. Synge, personal communication). There seems to be little in the way of a competent record of what is in fact eaten, or not eaten, by the food-gatherers of primitive, or even of sophisticated, human populations, and the basis on which particular species are selected or rejected.

Much more attention has, however, been paid to the question of the selection of particular species as suitable subjects for cultivation—the first step in sophistication. This subject has been especially dealt with recently by J. G. Hawkes (1967, 1969). An early step in the civilization of all cultures has been the cultivation and conservation of plant crops. "Most of the ancient Old World seed crops were domesticated in the mountains of the subtropics from about 25° to 45°N where these climatic conditions, with a cold winter, wet spring and autumn, and a hot, dry summer are to be found. The seeds needed to survive the long hot dry season in a well-baked thin soil, and there must consequently have been a strong selection pressure for large seeds with large food reserves to resist the drying out and grow quickly when the rains came again. In these soils and under these conditions nothing with small seeds would survive well, but nor would large perennial plants either, so these ecological weeds, the ancestors of our cultivated plants, were able to grow and survive under these special conditions without competition from trees or herbaceous perennials" (Hawkes, 1969).

Why do these same seed plants persist as virtually the *only* such plants presently cultivated? The answer is probably to be found well expressed by Kay (1970) in the parallel situation of domesticated animals (a situation we shall have occasion to consider later on). "Once a settled mode of life had been adopted in Neolithic times it may have been more profitable and certainly less effort to adapt existing farm animals to changing needs rather than attempt to domesticate new species."

These considerations apply even more forcibly at present, when species are actually going out of cultivation under the pressures of urbanization. In what seems to be the rapidly approaching limit, only those species of perishable products which can be profitably canned, frozen, or otherwise processed will be available to the urbanized human individual; any remaining freedom of choice of species will be available only to those with resources of their own to grow them.

Given such freedom of choice, what would be the basis on which man would select the plants to be eaten? Professor John Yudkin (1969) answers this question as follows: "There are properties of shape, colour, smell, taste and texture that cause a particular species to choose particular foods. For these properties I use the word palatability, perhaps in an excessively wide sense." In all these respects, and especially in regard to smell and taste, chemistry is deeply and intimately involved. "Animals live in worlds dominated by chemical communication systems" (Arnold and Hill, Chapter 6). Except for colour (which especially dominates the world of the human animal), information about the relevant chemistry of plant species is almost non-existent. Even an adequate descriptive language for odour, on which a presumptive formulation for chemistry might be based, is lacking (see Rohan, Chapter 5, this volume). However, there is reason to hope that this is on the way to being supplied (von Sydow, 1971).

Yudkin's view that the most palatable foods for man are meat and fruit is supported by the fact that his increasing affluence is associated with an increasing consumption of these foods. So far as plant organs are concerned, it is the qualities that make the best-liked fruits attractive—sweetness of taste and "sweetness" of aroma—that will be attractive, and those that give cause for rejection—sharpness, bitterness and astringency—which will be repellent. Vegetables with savoury flavours such as peas and asparagus are also highly acceptable, a transfer reaction from the pleasure experienced in eating meat. It is significant that these are eaten with added salt (as also are nuts) rather than with added sugar. Because of man's ambivalence, he is not a very suitable species to use in studies of food selection, but he has the advantage over any other experimental animal in that he can say *why* a particular food is agreeable or disagreeable.

Otherwise, the situation with other animals is in many ways similar to that of man. In the wild, there may be a wide, if not, in fact, a free choice of food plants, but there are restrictions imposed on each species by virtue of its particular physiology. An interesting account of the way in which these restrictions lead to equilibrium of wild herbivores within a given ecosystem is given in the paper by Kay (1970) already referred to. "A richly varied population of herbivores makes full use of the vegetation available, different species grazing it in succession through the seasons and at all levels, from the tree tops to the ground." Domestication leads to a narrowing of the plant species available, but the range is still wide for the grazing animal (see Arnold and Hill, Chapter 6, this volume). With pets the situation is similar to that of the urbanized human (Rofe and Anderson, 1970), and with battery-fed stock, animals in zoos and laboratory animals (Lane-Petter, 1970) the choice is reduced to zero.

The animal is not only urbanized but imprisoned. In these conditions, an animal which is strictly herbivorous in the wild may change its food habits so completely as to take to eating meat (Schaller, 1963).

B. PLEASANTNESS AND UNPLEASANTNESS

This section actually begins not with animals eating plants but with animals eating eggs. In the early days of the Low Temperature Research Station at Cambridge, Dr T. Moran, Head of the Animal Products Section, set up a panel for tasting shell eggs, and this came into good use during the second world war for the evaluation of imported dried eggs. Just after the war, my colleagues and I were approached by Dr H. B. Cott, a zoologist working on protective coloration in birds, who had discovered that the flesh of birds with conspicuous plumage was usually distasteful, whereas that of cryptically patterned species was usually palatable. He had found the same with eggs, and he asked us to work alongside his team of animal tasters with our own trained panel (Cott, 1949). The results are shown in Figs 1–4.

With few exceptions the human and the animal tasters recorded the same verdict for the different species. There was one thing, however, that the human tasters could do that the animals could not, and that was to say *why* they disliked those to which they gave a low score; and almost without exception it was because they were *bitter*. These results were important for many reasons, but especially for two which concern us:

(1) These three animals, an insectivore (hedgehog), a rodent (rat) and a primate (human) tasted bitterness in equal degree.
(2) Bitterness was equally repellent to all three species.

These results gave me confidence to think that human sensory experiences in taste and flavour can be presumed to be the same in other animals—unless proven otherwise; and that bitterness is a *universally* repellent character in foodstuffs. And if indeed this is the case, might not its antithesis, sweetness, be a universal—in fact, *the* universal—attractant?

Returning now to plants and the grazing, browsing, vegetarian animal, we may ask the question "What are the possible sources of information about food preferences?"

Firstly, there is an enormous amount of information scattered through the wildlife literature, but nowhere, so far as I know, gathered together and documented for its own sake. Secondly, a great deal of work has been, and is being, done on the palatability and nutritive value of herbage and forage crops for livestock, most of which seems to show how immensely difficult a problem it is to discover the answer to the questions what exactly do stock eat when they are grazing and what governs their selection.

As regards the information available in the wildlife literature, I can give two examples from my own casual reading. The first of these is from "The Flight of the Unicorns" by A. Shepherd (1965). This deals with the last years of the oryx in Southern Arabia, the reference in the title being to the unicorn-like appearance of the two-horned animal when seen in profile. It graphically describes the unending search for food in the arid desert: the young and tender shoots of *Tamarix* and the ghada bush in the spring, the succulent root parasites *Cynomorium* and *Phelipaea*, their favourite *Tribulus* (caltrops) and nussi,

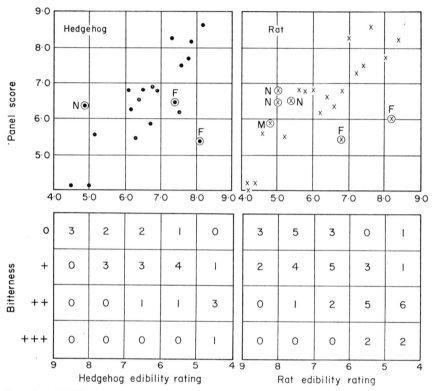

FIGS. 1–4. *Edibility rating:* order of preference of eggs exhibited by the animal. *Panel score:* average score for palatability given by trained human panel on a scale ranging from 0 to 10. Bitterness: score given by human taste panel. F(ruity), M(ealy) and N(utty) refer to flavour characters other than bitterness. Apart from these, there is a strong correlation between adverse score and bitterness

a sweet grass (*Aristida* sp.). Here the food intake is dictated by the need for it to supply water as well as nutrients.

A completely contrasting situation is described in another very much more detailed source, "The Mountain Gorilla" by George Schaller (1963). He lived with the gorillas in the uplands of the Congo for a year, recording their feeding and other intimate habits. They are mainly browsing animals, eating every day enormous amounts of leaves, stems and roots topped up with fruits and seeds. He gives a meticulous account of the plants and their parts that he observed them to be eating, and in many cases took the trouble to taste them himself and record their properties. Most remarkable of all, he distinguished bitterness from astringency—and this, we shall see later on, is a most valuable distinction.

"The first order of the day was to fill each colossal belly. There are many kinds of plants around Kimbala, but gorillas ignore all but 29 of them in their feeding. Their daily staples consist of bedstraw (*Galium*), wild celery, thistles, and nettles. When in season, bamboo shoots and the blue fruits of *Pygeum* (a kind of cherry laurel) may be taken. And when the mood takes them, they may bite the bark off several trees or the base of a sedge leaf. But all in all, they are not catholic in their tastes, and I could not determine why gorillas eat certain plants and ignore others. I found that, with few exceptions, the food plants tasted bitter or otherwise unpleasant to me."

His comments on the taste of the plants eaten are given in Table II. This Table is expanded from one given in his book (his Table 39) by the addition of the family to which the plant belongs, and of information on the presence or absence of tannins in that family, and some other repellent features of the plants concerned. The kind of detail with which the eating habits are recorded is illustrated by the following excerpt:

"*Galium sinense*. Three small rows of fairly blunt hooks grow on the back of each *Galium* leaf with the result that the vine readily adheres to fur and clothes, and, if pulled along the tender fleshy parts of the body, the leaves are quite abrasive. Gorillas handle the vine so as to neutralize the effect of the hooks, behaviour best illustrated by quoting from my field notes . . . she picks out several dry stems between thumb and index finger, using first one hand and then the other. Finally she pushes the *Galium* several times gainst her partly-opened lips while twisting the vegetation around in her hand, thus forming a tight green wad in which all leaves adhere to each other. She stuffs the mass into her mouth and chews."

The gorilla's unconcern with regard to the viciously stinging hairs of *Laportea* is remarkable. "The virulence of nettles at Albara was such that they readily burned through two layers of clothing; after a sojourn among them my knees were swollen and red welts covered my face. Yet gorillas handled them without hesitation and fed on stems and leaves that bristled with white hairs—the animals were apparently insensitive to them."

These are examples of devices developed by the plants as a protection from the browsing animal, but disregarded or circumvented by this particular one. Such devices are of two kinds: physical ones, such as these, and others such as thorns or leathery and dry, woody textures; and chemical ones of repellent taste and odour. It is these last that I want to deal with for the most part, because it is here that *specific* plant-animal interactions are to be found. The same is true also of attractants, some being concerned with textural properties such as tenderness and succulence, others with chemical properties such as sweetness of taste and "sweetness" of odour. It is the favourable balance between attraction and repellence which is the basis of selection.

TABLE II

Subjective evaluation of taste of some Gorilla food plants. From Schaller (1963); amplified.

Species	Family	Part eaten	Comments on taste	Tannin	Other characters
Arundinaria alpina	Gramineae	Shoot	At first relatively tasteless, but extremely bitter aftertaste	None	
Polypodium sp.	Polypodiaceae	Leaf	Slightly sour	Some	
Laportea alatipes	Urticaceae	Leaf	General leaf or grass flavour slightly astringent or bitter	Some	Intensely stinging hairs
Helichrysum sp.	Compositae	Leaf	Slightly bitter	None	
Rumex afromontanus	Polygonaceae	Inside of stem	Like garden pea pod	Some	
Peucedanum kerstenii	Umbelliferae	Inside of stem	Like radish, but bitter resinous aftertaste	None	Terpene
P. linderi	Umbelliferae	Inside of stem	Some like celery, some extremely bitter	None	
Carduus afromontanus	Compositae	Stem	Rather tasteless, slightly grass flavour	None	Prickles
Aframomum sp.	Zingiberaceae	Fruit	Some bitter and resinous, some sour and astringent	Some	
Cynoglossum geometricum and *C. amplifolium*	Boraginaceae	Root	Palatable, like radish	None	Hairy
Galium sinense	Rubiaceae	Whole vine	Like pea-pod and grass	None	Hairy
Droguetia iners	Urticaceae	Whole vine	Grass-like and slightly astringent	Some	?
Rubus runssorensis	Rosaceae	Fruit	Excellent, like domestic *Rubus*	Ellagitannin	
		Tender tips and leaves	Grass-like, slightly bitter and astringent	Ellagitannin	
Vernonia adolffrederici	Compositae	Pith	First tasteless, then bitter	None	
		Flowers	Slightly sweet	None	
Xymalos monospora	Uncertain	Fruit	Mealy	?	
Myrianthus arboreus	Moraceae	Fruit	Sour but excellent	?	
Erica arborea	Ericaceae	Bark	Slightly bitter	Some	
Hypericum lanceolatum	Hypericaceae	Bark	Bitter and astringent	Some?	Anthraquinone dyestuffs
Hagenia abyssinica	Rosaceae	Bark, pith	Bitter and astringent	Much	
Pygeum africanum	Rosaceae	Bark	Bitter, somewhat almond-like	Much	
Senecio erici-rosenii	Compositae	Pith	First tasteless then bitter	None	Alkaloids? Rank?

The terms used by Schaller (1963) in describing the unpleasant properties of the plants he tasted are fairly complete as regards coverage. The one he uses most frequently is bitterness. This is a property possessed by a large number of substances belonging to many different chemical classes, but outstandingly by alkaloids and cyanogenetic glucosides—strychnine (1) and amygdalin (2) can

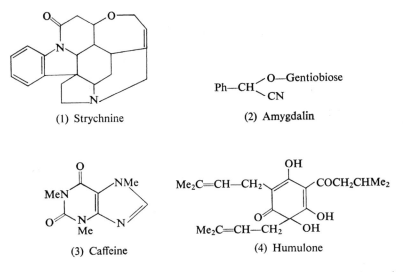

(1) Strychnine
(2) Amygdalin

(3) Caffeine
(4) Humulone

be quoted as familiar examples. That these happen to be extremely toxic is incidental—it is their bitterness which is repellent rather than the fact that they are deadly. Bitterness is also a relative term. Intensely bitter substances like strychnine (1) and the related brucine would render foods inedible at very low concentration, but other bitter substances such as caffeine (3), theobromine and hop bitters (e.g. humulone, 4) actually make food and drink more acceptable. There is no doubt also that familiarity leads to acceptability and even liking, so that a basically repellent character can become an attractive one. This may be the case with the gorilla's evident liking for plant tissues which the human taster recorded as unpleasantly bitter.

Next to bitterness, Schaller mentions astringency most frequently. This is not a taste, it is a "feel", and is associated with the presence of tannin in the food. Here again it is a question of the intensity of the sensation. A minor degree of astringency is attractive: red wine, an apple, or a cup of tea would be insipid if it were not to the expected degree astringent, but the sensation can soon become disagreeable if it is excessively evident. Many fruits are disagreeably astringent until they are "eating" ripe—pears and persimmons are well-known examples. The vegetative tissues of many plants are protected by the presence in them of potent tannins.

These are devices which seem, in their present environment, to have no other function in many plants but to act as deterrents to would-be predators. This seems to be the case also with the volatile constituents responsible for the offensive odours commonly described as "rank". These are of many kinds, but most often unsaturated hydrocarbons, alcohols and ketones of the terpene class such as those in wormwood (*Artemisia absinthium* L.). They are especially to be found in the Labiatae, the Umbelliferae and the Compositae. It is important to note here that some of these, such as citronellal (5) in lemon grass and geraniol (6) in roses may be sweet-smelling to us, but as repellent to many animals as turpentine or eucalyptus oil would be to us if present in foods. Citronella grass, *Cymbopogon afronardus* Stapf., for example was rated lowest of a dozen species of grasses grazed by buffaloes in Uganda.

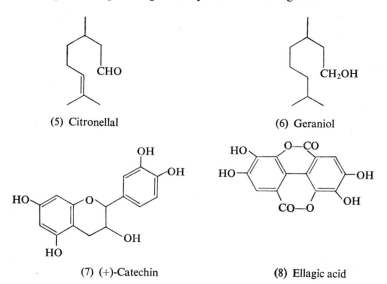

(5) Citronellal   (6) Geraniol

(7) (+)-Catechin   (8) Ellagic acid

It had occurred to me to give the title "Wormwood and Gall" to this contribution—but it would have involved an inexcusable pun on the two meanings of "gall". However, it would have drawn attention to the obnoxious nature of the major soluble constituent of oak galls—tannic acid. This is the best known of all vegetable tannins, and it may be surprising, therefore, to learn that it is really rather uncommon. Most tannins are either catechin tannins or ellagitannins, i.e. are structurally derived from either (+)-catechin (7) or ellagic acid (8). The former usually occur in the form of leucoanthocyanins. Either or both are regularly present in the tissues of woody plants, but are absent from those of herbaceous plants. Probably not one in a hundred of the plants growing in an English meadow will *have* tannins, while not one in a hundred of the trees in a tropical forest will *not* have them! It looks as if the change from the primitive arboreal habit to the herbaceous has been facilitated by the elimination of tannins, since this is found to be the case in so many angiospermous orders and families. But the loss of tannins for physiological reasons has resulted in the loss of astringency and the protection this affords against predators. So it is not surprising that other kinds of chemical protection are found in those plants from which tannins are absent—painfully pungent substances such as those in crucifers, onions and ginger, disgusting ones such as those in the Compositae and Verbenaceae, the bitter-almond odours and the bitter-tasting alkaloids. Plants that have tannins seldom have these other means of self-defence.

Qualitatively, it does not seem to matter whether the tannin is an ellagitannin or a catechin tannin: both are astringent and, above a certain level, reduce the palatability of the feed, and both impair the digestibility of the feed, especially of the protein in it. The two effects are, in fact, manifestations of one and the same property—the mutual precipitation of tannins and proteins: in the mouth, the precipitation of the proteins of the saliva and the mucous secretion of the buccal cavity; in the stomach or rumen, the precipitation of the proteins and enzymes and their withdrawal as nutrients and agents of digestion. To qualify as a tannin, however, it is not sufficient for the molecules to be present just as members of a particular chemical class, they must fall within a fairly narrow range of molecular size and complexity. If they are too small they will not precipitate the proteins; if they are too large they will be insoluble. So it is not sufficient to analyse foods for particular chemical groupings like catechol residues (e.g. using permanganate or Folin-Denis reagent, the common ways of estimating "tannin" content), nor even to determine the tanning properties of an extract, because the tannins and proteins may have precipitated each other by the very process of extraction. However, chemical analysis will provide useful information provided it is supported by actual determination of the palatability or digestibility of the food or feeding stuff concerned.

Of the two methods, the one chemical, the other physical, currently in use, the physical method is likely to provide information more closely related to the actual astringency of the tannin. Ideally, the closest to the real thing would be to mix the masticated food with saliva and measure the amount of salivary mucin precipitated. But saliva is neither easy to come by nor is the precipitate of mucin easy to measure. However, we have a source of coagulable protein always (literally) ready to hand in the form of finger blood, and this has proved to be simple and efficient to use.

As regards chemical methods, it is important to know just what it is one is trying to determine. Many of the standard methods determine particular chemical groupings which are not at all specific to tannins, and the relationship of the figures quoted are only fortuitously related to the amount of *functional* tannin present. However, this is on the way to being rectified, in particular by Robert E. Burns at Georgia Experiment Station. In forage crops there is seldom any tannin present other than catechin tannin, and this is almost always in the form of leucoanthocyanin. Burns (1963) has developed a method based on a well-known reaction for these compounds with vanillin, using as a standard a stock sample of *Sericea lespedeza*, a leguminous forage plant much grown in the southern United States but unfamiliar in England; it is closely related to sainfoin, which also has the same kind of tannin. Ellagitannins do not react, but there are specific reagents for these tannins also. We now have, therefore, means of determining fairly accurately the amount and kind of *chemical* tannin present in foods and feeding-stuffs, but this will not tell us how astringent and unpalatable these will be, nor how much their digestibility will be impaired.

The case of tannins illustrates what can be done in the way of relating chemical composition to systematic botany. So far as many other classes of constituents are concerned, "chemical plant taxonomy" is often the only way at the present time of guessing why some kinds of plants may be attractive and others repellent. "The kinds of substances that a plant contains depend upon the kind of plant that it is" (Bate-Smith, 1958). This kind of information is growing at a great rate, as evidenced by the increasing numbers of papers under the heading of Chemotaxonomy in the journal *Phytochemistry*. "Kinds of plants" are becoming increasingly capable of being chemically defined.

For this information to be usable, however, it is necessary for the subjective properties involved in attraction and repellence to be representable in chemical terms. That this is broadly possible is clear to anyone sufficiently familiar with the odours of chemicals—the fruity odours of esters, the pungency of mustard oils, the balsamic odours of many aromatic (i.e. benzenoid) compounds, but the present limitations in the cases of odour and pungency are the subject of the chapter by Rohan which follows this one.

REFERENCES

Bate-Smith, E. C. (1956). *Sci. Proc. Roy. Dublin Soc.* **27**, 365.
Burns, R. E. (1963). "Methods of Tannin Analysis for Forage Crop Evaluation." Georgia Experiment Station Tech. Bull. NS 32.
Cott, H. B. (1949). "The Palatability of the Eggs of Birds." *Oologist Record* (Reprint).
Harper, R., Bate-Smith, E. C. and Land, D. (1968). "Odour Description and Classification." Churchill, London.
Hawkes, J. G. (1967). "Crops, weeds and man." Inaugural Lecture, University of Birmingham.
Hawkes, J. G. (1969). *In* "The Domestication and Exploitation of Plants and Animals" (P. J. Ucko and G. W. Dimbleby, eds), p. 17. Duckworth, London.
Kay, R. N. B. (1970). *Proc. Nutr. Soc.* **29**, 271.
Lane-Petter, W. (1970). *Proc. Nutr. Soc.* **29**, 335.
Rofe, P. C. and Anderson, R. S. (1970). *Proc. Nutr. Soc.* **29**, 335.
Schaller, G. B. (1963). "The Mountain Gorilla." University of Chicago Press, Chicago and London.
Shepherd, A. (1965). "The Flight of the Unicorns." Elek Books, London.
Ucko, P. J. and Dimbleby, G. W. (1969). *In* "The Domestication and Exploitation of Plants and Animals." (P. J. Ucko and G. W. Dimbleby, eds.), p. xvii. Duckworth, London.
von Sydow, E. (1971). *Fd. Tech.* **25**, 40.
Williams, H. (1963). "Ceylon", 2nd edition. Robert Hale, London.
Yudkin, J. (1969). *In* "The domestication and exploitation of plants and animals" (P. J. Ucko and G. W. Dimbleby, eds), p. 547, Duckworth, London.

CHAPTER 5

The Chemistry of Flavour

T. A. ROHAN

Bush Boake Allen Ltd., Hackney, London

I. Introduction

It is regrettable that, whereas much is known about the flavour of chemicals, there is very little known of the basic principles governing the chemistry of flavour. It is unlikely, therefore, that we can yet formulate hypotheses to account for the palatable quality of the food we eat, and the position is complicated by the creation of a whole range of flavours unknown in nature. These are resultant upon preprandial processing, particularly heating which results in the formation of artefacts. In this respect, man has not only come to terms with his environment but has acquired the ability to modify the environment to suit his taste, in the most literal sense.

Before continuing, and in order to avoid confusion, it would be wise to define the expression "Flavour" as it is now generally accepted. Flavour comprises three principal characters described variously as tactual, gustatory and olfactory, or feel, taste and aroma.

The tactual component concerns the feel or texture of the food in the mouth (smooth, chewy, particulate, fluid, etc.); the gustatory component concerns the limited number of sensations which are detected on the tongue (salt, sweet, sour and bitter) and to these might be added astringency. The olfactory component involves the smell of the food, and is now separated into two distinct contributions; these are the odour, which is perceived when the food is smelled, and the aroma, which is sensed by the olfactory receptors whilst the food is being eaten.

Each of the five basic senses: sight, touch, smell, hearing and taste is involved in flavour appreciation. The first impression one has of food is its appearance which can either attract or repel. On subsequent passage to the mouth part of the volatile fraction is inhaled through the nostrils, thereby coming in contact with the olfactory receptors which are situated in the roof of the nasal cavity. When the food is ultimately chewed the basic taste and texture sensations are observed as it comes in contact with various areas of the tongue. At the same time the volatile aromatic constituents escape through the nasopharynx and come into contact with the olfactory receptors augmenting the sensations already perceived by smelling. The sense of hearing is called into play with foods like raw celery or some breakfast cereals.

Of all the senses involved, that of smell is the really important component of flavour. When the nose is out of action, as is a feature of the common cold, food loses its appeal and its character, and it may be difficult to recognize the food being eaten. The same phenomenon may be experienced by eating with the nose blocked by pinching between the thumb and forefinger.

Despite the paucity of knowledge of the chemical processes involved in the acceptance of foods there exists much information on the flavour characteristics of many chemical substances and their occurrence in foodstuffs. Research has shown that, although no single compound is totally responsible for the flavour of any one fruit, some organoleptically important constituents have been isolated: amyl esters from banana; citral from lemon; undecalactone from peaches and ethyl decadienoate from pears (for review, see Nursten, 1970).

Kovats (1963) found that mixtures of methyl anthranilate and thymol, in the correct proportions, have the characteristic aroma of mandarin oranges. The same author showed, by his work on lime flavour (Kugler and Kovats, 1963) what difficulties may be encountered in flavour research. Forty-four compounds were found which constituted 97·6% of the flavour volatile fraction, and the remaining 2·4% comprised at least 100 components in trace quantities. Lime is not unique, and the science of flavour chemistry is becoming more and more involved in the study of trace constituents whose contribution to the characteristic flavour of foods is critical.

Subjective testing is imperative at all stages in flavour research studies, if results are to have any meaning other than the bare chemical composition of foods. The work of Gold and Wilson (1963) on celery exemplifies both the application of sensory evaluation and the importance of constituents which are present in foods at trace levels. The flavour volatiles in celery, which were present at a level of about one ppm, were isolated by fractionating the expressed juice from five tons of the vegetable. Four alkylidene phthalides were identified which were found to be intimately associated with celery flavour. When these were mixed with two other compounds identified during this study (*cis*-hex-3-enyl-pyruvate and diacetyl), the flavour and aroma of celery were reproduced. The results were confirmed by taste panel and confer upon this work the distinction of demonstrating the importance of the use of organoleptic control in flavour research.

The eating habits of civilized man are largely governed by the flavour of the food he selects, and the long held belief that man will select those foods necessary for his physical well-being has been shown to be unsupportable. The hedonic qualities of food are of more importance, in determining what is eaten, than nutritional value, and the remarkable phenomenon of malnutrition in the so-called developed societies is consequently a reality.

II. The Anatomy and Physiology of Flavour

The taste buds are grouped in papillae on the surface of the tongue, and most papillae appear to be sensitive to more than one taste. There are regions of distribution of the four main kinds of taste receptor. The sweet taste is more easily sensed on the tip of the tongue, the bitter taste at the back, the sour taste at the edges, and the saltiness on the tip and at the edges of the tongue.

The olfactory receptors (numbering more than one million) are located in a small region at the top and towards the rear of the nose. They are tightly packed in this region and are protected from direct contact with the outside environment by means of a series of folds or turbinates.

Attached to each olfactory receptor is a number of short and long cilia, or hair-like filaments, which lie within a mucoid material bathing the receptor region. Some investigators have suggested that the receptor sites are associated with the cilia although others have minimized their role and have suggested instead that the receptors are located on the cell itself.

Whether it is the cilia or the cells which act as receptors is uncertain, but receptor cells connect directly with the olfactory bulb in the brain and are capable of transmitting some 10^8 bits of information per second, which is of the same order as that of the optic nerve system. It is not certain if the olfactory cells are mono-functional or if they respond to more than one stimulus, nor is there any information on the mechanism of the cell action.

Research in this field is comparatively new and is obviously extremely tedious. One approach to the problem of obtaining a greater understanding of the mechanism of odour perception is to insert electrodes into selected cells in the olfactory epitheleum of a newly dissected frog and to subject the system to the vapours of volatile organic chemicals. The electrical response in the cell resulting from stimulation with various types of organic materials is recorded, and by varying the position of the electrode it is possible that some light may be thrown on the mechanism of the cell action and perhaps some simple relationship between this action and the nature of the stimulus.

This type of approach, which is being employed by a colleague in Bush Boake Allen Ltd., might even lend itself to the study of the phenomenon of synergism whereby substances present at under threshold amounts may influence the flavour.

The aspect of flavour whereby one substance may influence profoundly the effect of another is well exemplified in the range of substances known as flavour potentiators. These are compounds which alter or augment the response of sensory organs to the chemical stimuli of foods, and do not necessarily of themselves impart specific flavour notes. Perhaps the most common flavour potentiator is salt. Sugar too has this quality and, when added in small quantities to some foodstuffs, acts as more than a sweetening agent.

Japanese food scientists who have made a special study of flavour potentiators, in 1909 isolated from Sea-tangle, monosodium glutamate (MSG). MSG has been produced commercially for 50 years and is generally accepted as an excellent flavour enhancer or potentiator.

The 5'-nucleotides have only recently been introduced as flavour potentiators although the first flavour nucleotide derivative was discovered in Japan in 1913 when a crystalline flavour compound isolated from dried bonito was reported to be the histidine salt of inosinic acid. It has only recently been established that the nucleotide element is the flavour factor and that histidine is irrelevant. Flavour potentiator activity was shown exclusively by 6-hydroxy-purine-5'-mononucleotide (1).

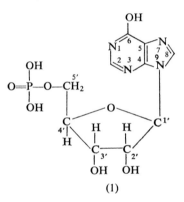

(1)

Another physiological phenomenon, which is far from understood, is that which is associated with chewing the berries of the African "Miracle fruit" (*Synsepalum dulcificum* Don, Sapindaceae), which have the property of eliminating sourness. After chewing these berries, lemons taste similar to sweet-orange. The substance responsible for this effect has been isolated and found to be a glycoprotein with a molecular weight of 44,000, and a mechanism has been presented to account for this phenomenon which relies upon physical binding of the protein to the taste membrane (Kovats, 1963; Brouwer et al., 1968; Kurihara and Beidler, 1968).

The anti-saccharine properties of the leaves of the tropical plant *Gymnema sylvestre* R. Br. (Asclepiadaceae) have been known for well over a century. After chewing one or two of these leaves, one is unable to detect sweetness, and bitterness is also partially suppressed (Stocklin, 1969). The active principle is acidic and glycosidic, and is a multicomponent mixture. The major component, gymnemic acid A, exhibits anti-sweetness properties.

Four gymnemic acids have been recognized and when the mixture is hydrolysed, the nature of the degradation products suggest that the original compounds are D-glucuronides of a hexahydroxyterpene (esterified with different acids) (Scheme 1).

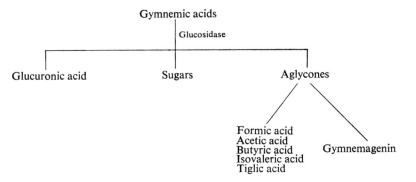

The structure of gymnemagenin has not been fully elucidated but is suspected to be the pentacyclic triterpene (2) (Stocklin, 1969).

These two examples of taste modifiers suggest that others may exist in the plant kingdom and, if readily available, could be as important in food chemistry as the flavour potentiator. A study of the mechanism of their action

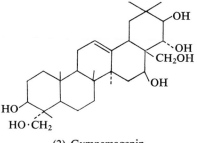

(2) Gymnemagenin

might even lead to the production of simpler synthetics whose function could be predicted and tailored to meet specific requirements.

III. The Language of Flavour

Unlike vision and audition, the stimulus for olfaction is chemical, and if we define the chemistry of flavour as the relationship between organoleptic sensation and the chemical nature of the stimulus, the reason for the comparatively slow rate of advance in the science of flavour chemistry becomes apparent.

Despite many attempts to correlate chemical structure with flavour sensation there have not yet emerged any guiding principles. This is not surprising when one reflects on the system used to depict the structure of organic compounds. The formulae employed are meaningful only as intellectual aids to presentation and discussion of the results of research, but as such are indispensable as a language for chemists.

Chemistry as a science is not dogmatic and simply utilizes means of communication to describe how certain phenomena might be explained. When attempting to relate the chemical constitution of a molecule to its physiological behaviour, difficulties arise if the chemistry of the process is not completely understood. This is true of the flavour response, and the use of the accepted intellectually convenient methods of drawing pictorial models in two dimensions can be very misleading.

Compounds exist which are of widely divergent molecular structure but similar in flavour and the converse is also true. Consider the two closely related lactones, undecalactone (3) and α-nonalactone (4); the former exhibits a peach odour and the latter a coconut odour.

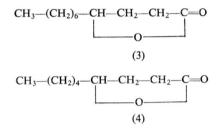

$$CH_3—(CH_2)_6—CH—CH_2—CH_2—C\!=\!O$$
$$\underline{\quad\quad\quad O \quad\quad\quad}$$

(3)

$$CH_3—(CH_2)_4—CH—CH_2—CH_2—C\!=\!O$$
$$\underline{\quad\quad\quad O \quad\quad\quad}$$

(4)

Attempts to classify flavours in a manner aimed at the establishment of primary members in a similar way to colours exemplify this difficulty. Aromatic substances have been classified into seven primary groups by Amoore (1962) and these are Camphoraceous, Pungent, Ethereal, Floral, Peppermint, Musky and Putrid. (For a more comprehensive survey of current systems of odour description, see Harper et al., 1968.)

These, and other terms, may be useful in attempting to derive a relationship between odour and objective characteristics, and Amoore (1962) has related odour quality empirically to the size, shape and electronic state of the molecule. However, to the flavour chemist who is bent on the isolation of a specific characteristic odour, this type of descriptive appraisal is of very limited value.

This can be appreciated if we consider the two soft fruits, strawberry and raspberry. The breadth of definition of the descriptive terms listed above is so wide that, only by using the terms "strawberry" and "raspberry" in his work, is the chemist likely to make progress. Our flavour vocabulary is not yet adequate to the role of unequivocal description, in general terms, of the flavour impact components of a foodstuff, nor is our knowledge of the relationship between the structure of a molecule and its flavour sufficiently great to permit prediction.

No one could have forecast that ethyl methylphenylglycidate (5), the ethyl ester of methylphenylglycidic acid, would have a strong strawberry odour, or that nona-2:6-dienal (6) would have an aroma identical with cucumber.

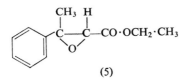

(5)

$$CH_3-CH_2-CH=CH-CH_2-CH_2-CH=CH-CHO$$
(6)

Is it fortuitous that after years of painstaking work on the volatiles of chocolate and coffee, research chemists have not succeeded in identifying any single compound characteristic of either of these common commodities, yet almost 700 compounds have been identified? Are these materials characterized by mixtures of compounds in proper balance or have the key compounds eluded us?

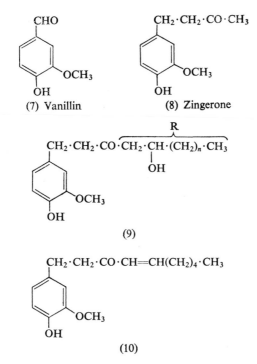

(7) Vanillin   (8) Zingerone

(9)

(10)

Who would have dared predict that the condensation of the carbonyl group in vanillin (7) with acetone could result in a complete change of character to provide the hot pungency of ginger, as in zingerone (8)?

Attempts to arrive at universal relationships between flavour and chemical structure have probably been over-ambitious and it might well be more rewarding to confine each study in this field to a much smaller and more clearly defined area. In this way it might be possible to relate the empirical approach with the more systematic one of attempting to unravel the physiology of the process.

As an example of flavour structure relationships, confined as suggested above, we may consider the pungent principles associated with any of the species: ginger, pepper and capsicum. Pungency is not a true flavour effect and is, in this context, perceived on the tongue and back of the throat. Nevertheless it is a very clearly defined sensation and a number of chemical compounds exhibiting this effect have been isolated and identified.

The principal pungent constituent of ginger is gingerol (9). Two other compounds have been isolated from ginger extracts which, although known to be artefacts, exhibit pungency. These are zingerone (8) and shogaol (10).

Subsequent to the demonstration by Connell (1969) that zingerone and shogaol are artefacts, the isolation of another pungent compound, paradol (11), was described. The side chain can vary in length in gingerols, shogaols and paradols, and it is of interest that pungency increases with the length of the side chain in synthetic shogaols up to $R = C_5H_{11}$.

The roots of the plant *Curcuma longa* L., a member of the Zingiberaceae,

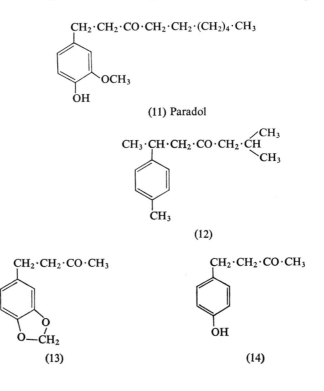

$$CH_2 \cdot CH_2 \cdot CO \cdot CH_2 \cdot CH_2 \cdot (CH_2)_4 \cdot CH_3$$

OCH$_3$

OH

(11) Paradol

$$CH_3 \cdot CH \cdot CH_2 \cdot CO \cdot CH_2 \cdot CH \begin{smallmatrix} CH_3 \\ CH_3 \end{smallmatrix}$$

CH$_3$

(12)

$$CH_2 \cdot CH_2 \cdot CO \cdot CH_3$$

O
O—CH$_2$

(13)

$$CH_2 \cdot CH_2 \cdot CO \cdot CH_3$$

OH

(14)

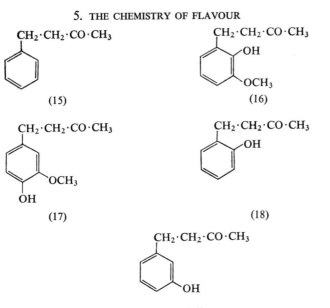

(15)

(16)

(17)

(18)

(19)

yield an essential oil, the main constituent of which is *ar*-turmerone (12). That *ar*-turmerone exhibits pungency is of some interest in view of its structure.

Nomura (1918) synthesized a number of compounds related to zingerone (8) but with variations either in the side chain or in the *aromatic* nucleus. Some of these compounds (13–19) are illustrated here and, as far as can be ascertained, Nomura found 4-hydroxy,3-methoxy substitution to give the greatest pungency.

Pravatoroff (1967) has made some general observations on pungency of zingerone and related compounds in which variations in structure are considered. These are as follows:

(1) One phenolic hydroxyl and one methoxyl group are necessary.
(2) Multiple hydroxyl groups reduce pungency.
(3) Increase in acid character of hydroxyl group by introducing electrophilic groups reduces pungency.
(4) Methylation of free hydroxyl groups reduces pungency.
(5) Replacing hydroxyl and methoxyl groups by a methylenedioxy group reduces pungency.
(6) Replacement of 4-hydroxy,3-methoxy groups by 4-hydroxy,2-methoxy or 4-methoxy,3-hydroxy groups reduces pungency.
(7) Reduction of side chain carbonyl to secondary alcohol reduces pungency.
(8) Increases in length of side chain to *N*-amyl strengthens pungency; further increase reduces pungency.

Pungency would therefore seem to be fairly closely linked with the structure (20).

4

There are other pungent principles which occur in spices, notably capsaicin (21) in capsicum and piperine (22) in black pepper. The double bond in the capsaicin side chain is not essential to pungency. Pepper is much less pungent than capsicum, which is in accordance with the hypothesis that pungency is reduced when the 4-hydroxy,3-methoxy system is replaced by a methylene-dioxy group.

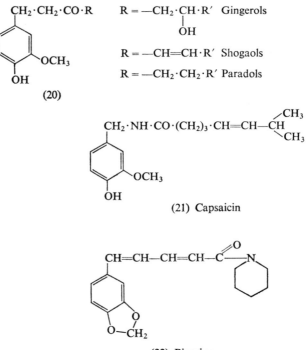

$CH_2 \cdot CH_2 \cdot CO \cdot R$

$R = -CH_2 \cdot CH \cdot R'$ Gingerols
             OH

$R = -CH=CH \cdot R'$ Shogaols

$R = -CH_2 \cdot CH_2 \cdot R'$ Paradols

(20)

$CH_2 \cdot NH \cdot CO \cdot (CH_2)_3 \cdot CH=CH-CH \overset{\diagup CH_3}{\diagdown CH_3}$

(21) Capsaicin

$CH=CH-CH=CH-C \overset{\diagup O}{\diagdown N}$

(22) Piperine

This is a relatively restricted example of the way in which the chemistry of a group of substances may be related to its physiological effect. Nevertheless, it is a positive, though modest, contribution and it is tempting to hope that other pungent commodities will be studied; and that the data available for physiologists will be multiplied.

Basic research on natural flavours has not been as unproductive as might be inferred from the earlier statement that there is no recorded example of any such flavour being totally separated into each of its components which, after identification, have been reassembled successfully. Frequently, it is sufficient only to identify the constituent responsible for the primary aroma of flavour. Indeed even were the volatiles totally identified, it is unlikely that the original flavour could be faithfully reproduced without a knowledge of the nature of the non-volatile residue.

IV. TECHNIQUES IN FLAVOUR RESEARCH

Invariably in flavour research, the first task is the formation of a taste panel incorporating one trained flavourist, to assist the chemist in describing, in common terms, the aroma and/or taste of his raw material. This is the product which the creative flavourist is attempting to simulate.

Having defined the problem in terms which have been arrived at by the taste panel, the chemist must then produce an extract or concentrate which comprises the required flavour characteristics. This may involve solvent extraction, ordinary distillation, vacuum distillation, steam distillation or vacuum steam distillation; or even entrainment in a stream of gas, and is the most critical stage of the operation. Frequently very large amounts of material must be processed to obtain relatively small amounts of isolate. McFadden and his colleagues (1965) isolated 50 grams of oil from ten tons of strawberries, i.e. 5 ppm; when one considers that this oil comprises a large number of constituents, some present in trace amounts, the problem begins to assume major proportions.

Another approach is trapping and subsequent analysis of the head space vapours from a single fruit and this represents the other end of the scale, in which small amounts of raw material are used to give micro and submicro amounts of volatile materials for analysis.

The isolate must also be subjected to sensory evaluation to ensure that it embodies those qualities which have been already included in the definition of the problem. It is at this stage that much of the earlier work on flavour chemistry understandably failed. It is extremely difficult to conduct this type of subjective evaluation and, as is so often true, when time is limited, such work is neglected. The growth of Food Science Departments in many Universities has done much to correct this defect, and there has been a growing awareness that flavour chemistry cannot be divorced from sensory evaluation and, if it is, the results are no real contribution to this field of knowledge. Emily Wick (1960) has put it in unequivocal terms that "the best friends a chemist can have, at the beginning of a flavour investigation, are a group of people wise in the attributes and characteristics of the flavour in question, who can, by organoleptic evaluation, help him to make the basic decision as to whether the non-volatile, or the volatile fraction, or both, must be studied."

After sensory evaluation, the concentrate is now freed from non-volatile components of the fluid and is consequently but a shadow of the starting material. Nevertheless, it should contain those compounds which characterize the food in question. The final stage in the sequence is the fractionation of the essence into broad cuts which are again subject to sensory evaluation and only that fraction which comprises the desired notes is selected for further study.

The process is continued until a fraction of relatively simple composition is obtained. If sufficient of this fraction is available it may be resolved further by preparative gas chromatography or countercurrent extraction and individual constituents identified. Alternatively, it may be found necessary, because of the small amount of sample available, to attempt simultaneous resolution and identification of the various components by combined gas chromatography—mass spectrometry.

It has been stated erroneously that the final requirement is that a complete breakdown of the volatile constituents of a flavour should culminate in the reconstitution of the original flavour. This is indeed an extravagant objective which, although always the ultimate objective, is loaded with excessive optimism for present day knowledge and techniques. Something short of this, and somewhat simpler to achieve, is more practical and may yield commercially rewarding results.

Modern sophisticated instruments are capable of handling very small amounts of material and the gas liquid chromatograph can detect the presence of as little as 1 nanogram (10^{-9} g) in a sample. Despite the advances made in gas chromatography this instrument is, at best, an effective means of resolving a mixture of volatile substances into its constituents. The over-interpretation of such data has been a major problem in flavour chemistry and has resulted in the publication of long lists of compounds whose presence in a foodstuff is no more than suspect.

It can be demonstrated that the retention on a gas chromatography column of a given compound in a complex mixture may be markedly influenced not only by its concentration in the mixture but also by the concentrations of other components. Retention times can be meaningful if they are determined on isolated compounds on several different columns whose characteristics remain constant and using different sized injections which permit extrapolation to zero concentration. Unless one is dealing with a relatively simple mixture of reasonably constant proportions, retentions should not normally be regarded as satisfactory criteria for identification.

Finally, when examining gas chromatograms it should be remembered that the detector is not by any means a true facsimile of the nose and that very large peaks on the chromatogram may have very little aroma and vice versa.

V. Summary and Conclusions

The current state of fundamental knowledge in the field of flavour chemistry is not very great, considering the effort which has been expended in the past twenty years or so.

This can be ascribed to the almost total lack of a means of communication, a language of flavour, and to the failure of many chemists in the past, to appreciate the necessity for sensory evaluation. Concerning the means of communication, consider how aware we are of flavours of strawberry, cloves, ginger, cheese and chocolate, yet how would these flavour sensations be described to someone who had not experienced them?

Sensory panels are an essential part of the equipment of the flavour research chemist who should not fall into the trap of assuming that his technology alone is sufficient to enable him to solve problems in this field. Flavour research is a field for co-operation of chemist, sensory analyst, expert flavourist and synthetic organic chemist, preferably working in a team or in close harmony. Co-operative effort is likely to yield practical results in the short term but we will never really interpret the results of our work, or arrive at a position where we can predict the flavour of a new compound, until the physiology of this process is fully understood.

REFERENCES

Amoore, J. E. (1962). *Proc. Sci. Sec. Toilet Goods Assoc., Suppl.* **37**, 1.
Brouwer, J. N., van der Wel, H., Francke, A. and Henning, G. J. (1968). *Nature, Lond.* **220**, 373.
Connell, D. W. (1969). *Food. Tech. Austral.* **21**, (2), 570.
Gold, H. J. and Wilson, C. W. (1963). *J. Fd. Sci.* **28**, 484.
Harper, R., Bate-Smith, E. C. and Land, D. (1968). "Odour Description and Classification." Churchill, London.
Kovats, E. (1963). *Helv. Chim. Acta* **46**, 2705.
Kurihara, K. and Beidler, L. M. (1968). *Science, N.Y.* **161**, 1241.
Kugler, E. and Kovats, E. (1963). *Helv. Chim. Acta* **46**, 1480.
McFadden, W. H., Teranishi, R., Corse, J., Black, D. R. and Mon, T. R. (1965). *J. Chromatog.* **18**, 10.
Nomura, H. (1918). *Sci. Rept. Tokyo Univ.* **7**, 67.
Nursten, H. E. (1970). *In* "The Biochemistry of Fruits and Their Products" (A. C. Hulme, ed.), Vol. I, pp. 239–268. Academic Press, London and New York.
Pravatoroff, N. (1967). *Manfg. Chem. Aerosol News.* **38**, 40.
Stocklin, W. (1969). *J. Agr. Fd. Chem,* **17**, 704.
Wick, E. (1960). *Cereal Sci. Today* **5** (8), 240.

CHAPTER 6

Chemical Factors Affecting Selection of Food Plants by Ruminants

G. W. ARNOLD AND J. L. HILL

Division of Plant Industry, CSIRO, Floreat Park,
Western Australia

I. Biological Basis of Food Selection

There is a continuous flow of publications describing the food preferences of various ruminants grazing on a wide variety of plant pastures; but rarely does the work reported give any suggestion why particular plants or plant parts were grazed.

There have been several reviews in the last ten years in which the multiplicity of factors influencing food choice by ruminants are discussed (Arnold, 1964a,b, 1970; Gordon, 1970; Heady, 1964). Although attention is focused in this review on the chemical factors influencing food choice, it must not be forgotten that physical factors may considerably modify the responses of ruminants to the taste and odour of plants. For example, the location of the plant is important. A plant that has a highly attractive smell is much less likely to be eaten if it is underneath a prickly thistle than if it is growing surrounded by others of its own species. Also, despite an attractive odour, the feel to the lips and mouth and ease of harvesting and swallowing of a plant may be important in determining how readily it is eaten.

Animals, and ruminants are no exception, live in worlds linked by chemical communication systems. Their social, reproductive and feeding behaviour are primarily determined by chemical stimuli. The chemical signals which mainly influence food selection are those received at receptor sites for taste and smell. Stimuli are transmitted to the brain and the animal responds behaviourally or physiologically to the messages they contain. The response of the animal will be to integrate these messages with others, such as feedbacks telling of its current nutritional state or the presence of some metabolic disturbance. The desire to eat is essential before any intake requirements can be met. A hungry animal may respond by lowering either taste or smell rejection thresholds (Goatcher and Church, 1970).

The only signals that can activate this system are molecules which react chemically on receptors to transmit information. Therefore, in looking at food selection in relation to chemical composition of plants, only molecular concentrations of specific chemical forms are of significance. This hypothesis of communication implies that animals are able to distinguish significant stimuli from a background of chemical "noise". Thus, responses are given by the receptors only when receiving stimuli which are significant in that particular environment. What will be "noise" to one species could be key information to another, through either a differing sensitivity or a completely different response. The relative intensity of a signal will be determined not only by the number of molecules received, but also by the number of receptors that can receive them. Differences probably occur between and within animal species not only in the total number of receptors but also in the ratio of, say, bitter to sweet taste receptors. Thus, different animals of the same species might be expected to differ in their taste response in given situations. It is also quite probable that, for instance, a chemical which is sweet to one animal species may produce no response or a totally different response in another, due to a different molecular structural requirement or sensing mechanism at the sensory site.

It is only those chemical components of the plant that make contact with the senses that can influence food selection. This is the basis on which this review on plant chemical composition and food selection has been compiled.

II. THE INTERRELATED ROLE OF THE FIVE SENSES

In mammals the special senses of sight, hearing, touch, smell and taste are involved in food seeking and in food selection. The relative importance of the senses varies with animal species. In sheep, and this probably applies to other ruminants, the senses of touch, smell and taste are of greatest importance (Arnold, 1966a). The role of sight is primarily one of orientation to the flock and to the vegetation when animals are grazing, but there is no evidence that it is used in food selection (Arnold, 1966b). Although recognition of certain plants occurs (food seeking), selection of the parts of those plants to be eaten is by response to touch, smell and taste. There is no conclusive proof that sight is not used to select food, since little work has been done on this aspect.

In a series of studies with surgically treated sheep having single and multiple sensory impairment, Arnold (1966a) showed that touch, smell and taste were all important in food selection; marked changes in acceptability of plant species or strains of plant species were seen when each of these senses was impaired in turn. Not only is selection from a range of choices altered by impairing these senses, but total food intake may be increased or decreased.

In this work it was clear that only rarely was response to one of the senses of paramount importance in determining food preferences. There are two reasons for this: firstly, there is the variability between individual sheep in responses for each sense. Some sheep may be less sensitive to the coarseness of the leaves of a species than others, but more sensitive to its highly attractive smell than the others. Secondly, it is biologically more likely that choice of food is determined by the combined responses to several stimuli rather than to a single source of stimulus. Although touch is important, this review concentrates on the chemical senses.

III. RESPONSES OF RUMINANTS TO THE TASTE AND SMELL OF CHEMICAL SOLUTIONS

A. TECHNIQUE LIMITATIONS

The separation of the taste and odour responses of animals has rarely been examined. All the classical "taste" response behavioural studies really show responses to the "flavour" of the solutions tested, i.e. their taste and smell, and this may be complicated further by the temperature of the solutions. Rarely are these studies done under constant temperatures, although the

4*

chemical forms in the solutions used can be shown to vary with temperature. Baryshnikov and Kokorina (1965) report that optimum temperatures for cattle taste responses are 10 to 40°C. From 0 to −5°C taste discrimination diminishes greatly, mostly for bitterness and least for sweetness.

The roles of taste and smell relative to "flavour" could be expected to change with temperature-induced alterations in partial pressures and solubilities since both taste and smell depend on the presence of molecules within the mucus membranes covering the sensing areas; the effect of temperature at this interface could be substantial.

Before considering taste and odour responses in ruminants in detail, it must be stressed, *vide* Kare (1969), that different species live in distinct "taste" worlds which may or may not overlap. Certainly, the "taste" reaction of one species to a chemical stimulus is not a reliable guide to predicting the "taste" response of a second species. Likewise, the taste response of one individual of a species is not a reliable guide to predicting that of another individual in the same population. This certainly applies with ruminants. Troelsen *et al.* (1968) found that human assessments of sensory values of hays were not correlated with the voluntary intake of the hays by sheep.

It is necessary to explore what is known about responses of ruminants to pair choices of water and chemical solution in the classical taste response experiments, and their electrophysiological responses to chemical solutions applied to the tongue, before considering what ruminants taste in field situations.

B. ELECTROPHYSIOLOGICAL RESPONSES

Bernard and Kare (1961) studied electrophysiological responses in calves. The lowest concentrations to produce stimuli were 1 mM for NaCl, HCl and acetic acid, 10 mM for KCl and propionic acid, between 500 to 1000 mM for fructose, glucose and xylose, with quinine sulphate giving weak responses at 1 mM. They noted that there was no precise relationship between electrophysiological and behavioural responses. This is important because behavioural responses to quinine, for example, occur at concentrations of about 0·1 mM (see Fig. 1).

Baldwin *et al.* (1968) reported briefly on electrical responses of the chordatympani and glossopharyngeal nerves of the sheep, goat and calf. They state that in all three species there are strong reactions to acid and salt solutions and less response to quinine solutions; they found it difficult to elicit reactions to sugars by simple irrigation of the tongue. However, no indication is given of concentrations used.

C. BEHAVIOURAL RESPONSES

Bell (1959) with goats, Bell and Williams (1959) with calves, Stubbs and Kare (1958) with cattle, and Goatcher and Church (1970a,b,c,d) with sheep, cattle, goats and pygmy goats and Crawford (1970) with Columbian black-tailed deer

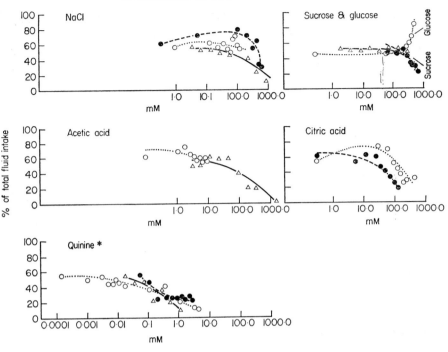

Solution concentration

FIG. 1. Taste response curves of sheep. Mean—ascending and descending, Arnold and Boundy, ●----●; Mean—ascending, Arnold and Boundy, o······o; Mean—ascending, Goatcher and Church, △———△ (1970c,d). Data are from three sources. Although curves were fitted for each set, those shown are drawn freehand to try and show probable curves for wider ranges of concentrations than used in any single experiment

* Quinine: Arnold and Boundy used quinine sulphate, Goatcher and Church used quinine monohydrochloride.

have studied behavioural responses by the "two choice" test method. Various thresholds have been proposed to describe responses obtained. The generally accepted thresholds, expressed as percentage of total fluid intake (TFI) taken, are 40 to 60% = neutral zone, 40% = rejection threshold and 60% = the acceptance threshold; 20% is taken as the threshold of strong aversion and 80% as that of strong preference.

In Fig. 1 the results for sheep obtained by Goatcher and Church (1970c,d) and by Arnold and Boundy (unpublished) are presented. Both research teams used several groups of sheep. This is important because of the between-animal variability in response which will be discussed later. A major problem with this method of studying taste response is that different response curves are obtained when tests are made in series with ascending concentrations than with descending concentrations (Kare and Ficken, 1963). In general, a higher percentage of TFI is obtained at a given solution concentration when ascending concentrations are used. Rumen metabolism and excretory mechanisms may be better adapted to an ascending series.

In Fig. 1, the curves from data of Arnold and Boundy show a mean curve for ascending and descending series, and curves for separate ascending series. Those of Goatcher and Church are for ascending series. Table I gives rejection

TABLE I

Approximate mM concentrations above which a solution is <40% of TFI

	NaCl	Glucose	Sucrose	Acetic acid	Citric acid	Quinine
Arnold and Boundy (asc + desc)	500	/	363	/	75	0·20*
Arnold and Boundy (asc)	/	>533	/	/	100	0·30*
Goatcher and Church (asc)	210	/	410	28	/	0·09†

* Quinine sulphate.
† Quinine hydrochloride.

thresholds calculated from equations describing the responses; there is approximate agreement between the three sets of data, except for quinine where different salts were used. Also, curves for glucose and sucrose are quite distinct. It must be stressed that because of the between group variability the standard errors for these thresholds, and also for acceptance thresholds, are large. Acceptance thresholds, where they have been found, vary considerably (Table II).

TABLE II

Approximate mM concentrations below which a solution is >60% of TFI

	NaCl	Glucose	Sucrose	Acetic acid	Citric acid	Quinine
Arnold and Boundy (asc + desc)	400	/	<73	/	<0·30	<0·05*
Arnold and Boundy (asc)	<0·9	334	/	4·0	<0·30	<0·00015*
Goatcher and Church (asc)	<27	/	<143	3·4	/	<0.016†

* Quinine sulphate.
† Quinine hydrochloride.

Either sheep dislike solutions of compounds used in the studies or like them only at very low concentrations.

Just how helpful are these studies of the basic modalities of taste in understanding selection of food plants? At present, very little, because chemicals like quinine sulphate and acetic acid are rarely found in plants, and even then are unlikely to be in the chemical form tested; further, the "taste" choice confronting the animal is not a simple choice situation but will involve levels of bitterness + levels of acidity + levels of sweetness + levels of saltiness mixed together. Hironaka and Bailey (1968) found that the voluntary salt intake of cattle could be increased by mixing sugar with salt; the presence of seventy per cent sugar was required to produce a significant increase in salt consumption.

IV. Factors Influencing Taste Preferences

A. EXPOSURE DURATION AND FREQUENCY

With classical methods, the length of time the animal is offered each solution at a given concentration could be expected to influence the results, i.e. the feedback information on differences in body loading of the solute may require

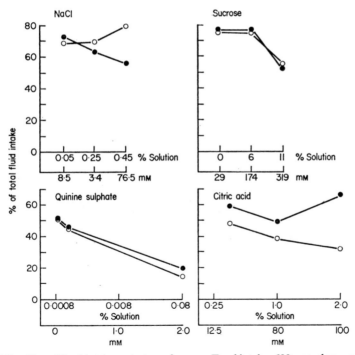

FIG. 2. The effect of food intake on taste preferences. Food intake: 600 g per day = o; 1800 g per day = ●. Four levels of intake were used in this study by Arnold and Boundy (unpublished). Levels of 1000 g and 1400 g per day gave responses intermediate to those shown. The test period was 96 hours

some time (days ?) before alterations in thresholds are produced. Thus it could be postulated (in the absence of evidence) that some rejection thresholds' concentrations will decrease for chemicals accumulating within the animal during increased length of exposure.

Number of exposures to a substance does not affect response (Arnold, unpublished) for any of the basic modalities. However, level of food intake strikingly affects response to citric acid and sodium chloride (Fig. 2). Whereas no acceptance threshold is found for citric acid when sheep are fed *ad libitum*, those on a restricted ration have an acceptance threshold at a concentration well above the rejection threshold for *ad libitum* fed sheep.

B. INDIVIDUAL ANIMAL DIFFERENCES

There is great variability in response within individual animals (Fig. 3). Such variability, in terms of the field selection of food plants, could have considerable ecological significance. It would mean that the individual would

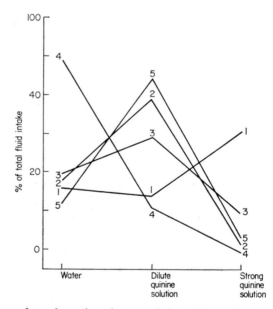

FIG. 3. Preferences of one sheep given the same choice of three solutions on five occasions. The five occasions were spaced over four months. Choice was over 24 hours. The variability shown was similar for each of another five sheep and for choices involving glucose, sodium chloride and acetic acid

not be restricted in its food selection by rigid preferences and aversions, so that the choice of food effectively available is large. When this variability is considered over a population of animals, e.g. a flock of sheep, its significance could be even greater. Nevertheless, both flocks of sheep and herds of cattle are highly selective when grazing on many plant communities which indicates that many of the plant species present elicit aversion responses from all individuals.

C. BREED DIFFERENCES WITHIN ANIMAL SPECIES

Despite the big variability between individuals, and between groups of animals of a species, there still remains sufficient genetic variability between breeds

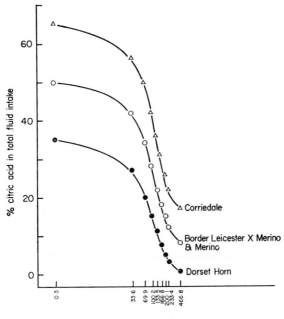

FIG. 4. Differences in preference curves of four breeds of sheep for citric acid solutions. Data are from Arnold and Boundy (unpublished). Tests were on groups of four sheep for 96 hours. The three curves differ significantly

of sheep to cause differences in taste responses. If no breed differences occurred, response curves would show similar TFI proportions for a given solute. Not only do proportions differ, but TFI and total solute consumptions also differ between breeds. Arnold and Boundy (unpublished) measured taste responses to sodium chloride, quinine dihydrochloride, glucose, acetic acid, citric acid, glutamic acid and tannic acid in four breeds of sheep. The response curves showing the proportions of TFI taken as chemical solutions, differed significantly only for citric acid (Fig. 4) and acetic acid. For these two substances, the differences were large. The amount of the solutes consumed differed substantially for quinine dihydrochloride and for acetic and citric acids (Fig. 5).

Total fluid consumptions are also found to differ significantly between breeds for a given solute intake. These relationships have not been considered previously in publications on taste responses. The ability or desire of sheep of different breeds to sustain these differences is of interest as the physiological reasons for these differences are unknown.

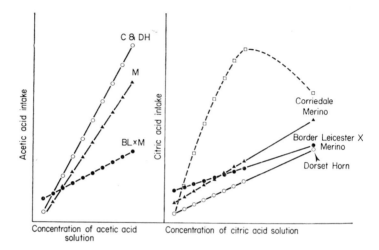

FIG. 5. Differences in voluntary intakes of citric and acetic acids amongst sheep of four breeds. Data are from Arnold and Boundy (unpublished). Tests were on groups of four sheep for 96 hours. For citric acid Corriedale sheep differ significantly from sheep of the three other breeds. For acetic acid the three curves differ significantly

D. ANIMAL SPECIES DIFFERENCES

Goatcher and Church (1970c,d) have made the most extensive comparisons between ruminant species. Sensitivity to chemical solutions, based on the lowest concentration to be discriminated, was as follows:

Sweet   cattle > normal goats > pygmy goats > sheep
Salty   cattle > pygmy goats > normal goats > sheep
Sour   cattle > pygmy goats = sheep > normal goats
Bitter   pygmy and normal goats > sheep > cattle

If some other threshold is taken, the results are different. For example, the molarities at which solutions are rejected (<40% TFI) rank, over animal species, as follows:

Salty   cattle > sheep > normal goats > pygmy goats
Sour   cattle > sheep > normal goats > pygmy goats
Bitter   sheep and cattle > normal and pygmy goats
Sweet   no rejection thresholds found

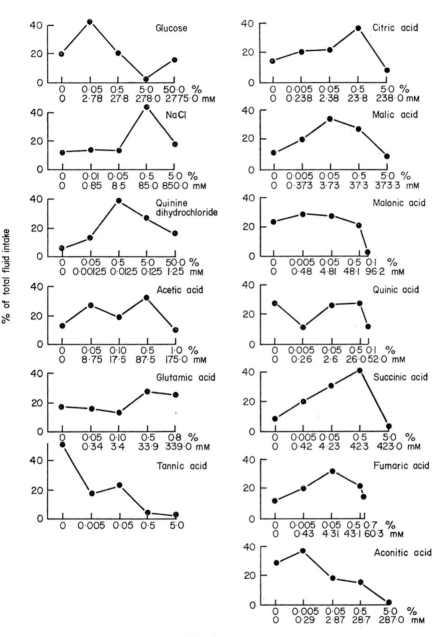

Solution Concentration

FIG. 6. Preferences of sheep offered five solutions of different concentrations of 13 chemical compounds. Data are from Arnold and Boundy (unpublished). Tests were on two groups of four sheep for 48 hours

Within each species the ranking of initial discrimination was as follows:

pygmy goats   sweet < salty < sour < bitter
normal goats   sweet < salty < sour < bitter
sheep   salty < sweet < sour < bitter
cattle   sweet < salty < sour < bitter

In another paper (Goatcher *et al.*, 1970) the above results are compared with those of Crawford (1970) for Columbian black-tailed deer. The main difference between these deer and the domestic ruminants was that the former had a great preference for low concentrations of sucrose solutions. The significance of these differences in the field situation has yet to be investigated.

E. SMELL INTERACTION

All the taste responses presented so far are for normal animals and the responses will thus include smell components. The comparative taste responses of normal and anosmic sheep, shown in Fig. 6 (Arnold and Boundy, unpublished) allow some assessment of the smell component. Quinine dihydrochloride, sodium chloride and glucose solutions appear to lack odours that influence taste responses, whereas the taste responses for most acid solutions are strongly influenced by their odour. The acids used were chosen because they occur in plants and may well have sufficient vapour pressure to be smelt by grazing animals.

F. MULTI-CHOICE PREFERENCES

As stressed earlier, the grazing animal often has a multi-choice situation, i.e. the plants present may vary greatly in concentration of particular chemicals. The response curves obtained from sheep given a choice of different concentrations of thirteen chemicals in solution are shown in Fig. 7. Response curves are obtained for almost every chemical and it is striking that in none of them is any of the choices totally rejected. This result has relevance to the field situation because both sheep and cattle, given the choice of many strains of a plant species, relatively rarely reject totally any of those on offer; where rejection does occur, it can usually be attributed to an obnoxious odour.

V. RESPONSES TO SMELL

There is very little information on this topic for ruminants, presumably because the experimental techniques are expensive and complex.

It has already been established that the sense of smell is critically involved in food selection by ruminants. Most of the work has been with sheep to determine which classes of compounds in food plants are involved.

FIG. 7. Taste response curves for normal and anosmic sheep. Data are from Arnold and Boundy (unpublished). Tests were on groups of four sheep for 96 hours

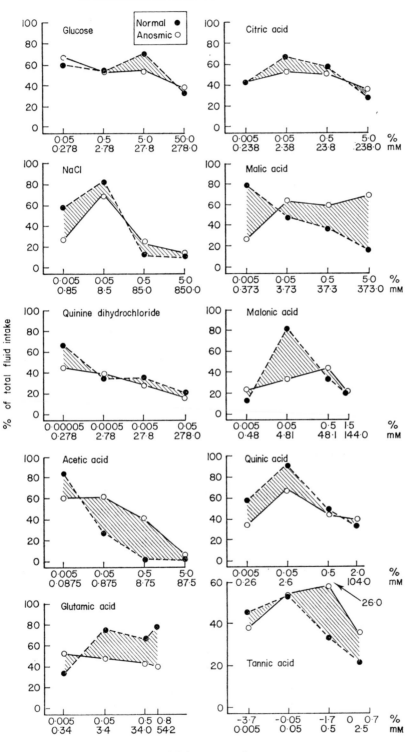

Solution concentration

A. ODOUR ADAPTATION

Tribe (1949) compared the preferences of normal and anosmic lambs for various pasture species and mixtures, and also amongst pasture treated with various volatiles. Although initially anosmic sheep more readily consumed herbage contaminated with what were considered obnoxious odours (to man), consumption by normal sheep occurred within quite short time periods (1 to 6 hours). Tribe concluded that adaptation to odours occurs rapidly and that smell can only be of supplementary importance in influencing the food selection of grazing sheep. The problem, as with taste studies in pens, is translation to multi-choice field situations.

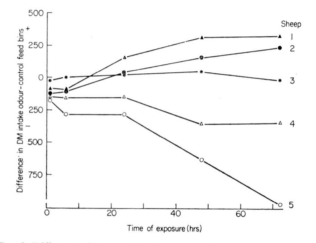

FIG. 8. Differences between individual sheep in response to odours

Arnold, Boundy and Knight (unpublished) used a choice between two food bins to test responses of sheep to odours. The odours were analytical grade chemicals soaked into pads of cotton wool and placed in the feed bins. A grass hay was used as the diet and uncontaminated bins had cotton wool pads soaked in water. Thirty-two odours were tested, and amounts of food eaten from the bins was recorded at 1, 6, 24, 48 and 72 hours. As with taste, so with smell; the between sheep variability was large. For 5 sheep, the proportion of food eaten from a bin had to be <0·32 or >0·67 for significance at the 1% level of probability. This again has ecological significance.

It is of interest that for a given odour a sheep gave a definite time response curve, and the curve differed for individuals. In addition, there was an overall pattern for individuals in their responses to all the odours. This is shown in Fig. 8. Most sheep appeared to take up to 24 hours to decide their response and then they maintained the response over the subsequent 48 hours.

The chemicals tested were:

6   Amino acids
6   Terpenes
11 Aliphatic alcohols, aldehydes, ketones and esters
5   Fatty acids
    Phenol
    Ammonia

Significant responses were obtained to only one amino acid (glycine), one fatty acid (butyric acid), and one alcohol (hexanol).

Kudryavtsev (1962) reported, but with no quantitative details, that Russian workers had shown that cattle differentiated between ammonia, 5% acetic acid solution, amyl acetate, camphor, bergamot, and lavender and cinnamon oils.

B. ROLE IN SELECTIVE GRAZING

Just how the sense of smell operates in selective grazing is not clear. We hypothesize that ruminants can discriminate odours only at short range and that they are thus able to make precise discriminations between adjacent plants and plant parts. This hypothesis is based on our own experimental evidence including an important observation noted with cattle. Diet samples were being collected from the cardia of rumen fistulated cattle grazing pasture which had many "dung patches". That the sense of smell was used by these cattle was established because of the marked auditory variation in inhalation intensity as the animals grazed. It was observed that deep inhalations were made when the muzzle passed over "dung patches" and these were left uneaten, but immediately a "dung patch" had been passed inhalations were no longer audible and the cattle grazed. In sheep, in the social bonding of ewes with their lambs, we have found also that odour discrimination only operates over very short distances.

VI. OTHER MECHANISMS INFLUENCING FOOD SELECTION AND INTAKE

A. INTAKE IN THE NO-CHOICE SITUATION

Although this paper reviews the selection of food plants by ruminants, it is of practical significance to know whether food intake, given no choice, can be enhanced or depressed by its taste and/or smell. That responses to both occur under grazing was shown by Arnold (1966a) using surgically prepared sheep. This is also true in pens using specific contaminants (Arnold, 1970); results are given in Table III. Nine odour and eight taste and/or odour contaminants were tested. The odours were provided by the cotton wool pad method and the taste/odour contaminants by addition to the rations. The contaminants chosen

TABLE III

Percentage change in intake due to contamination

Contaminant			
In feed	2·0%	Glycine	−25.2‡
	1·5%	Malonic acid	−19·3‡
	+5·0%	Tannic acid	−30·8‡
	+1·0%	Gramine	−61·5‡
	+1·5%	Coumarin	−32·2‡
As odours		Butyric acid	+10·6†
		Amyl acetate	+8·0†
		Hexan-3-ol	−8·5*
		Coumarin	−9·3†
		Glycine	−17·8†

* $= P < 0.05$.
† $= P < 0.01$.
‡ $= P < 0.001$.

Five sheep fed for 72 hours on control and contaminated feeds. Control feed was grass hay.

+ Physiological component involved since diet digestibility was depressed.

had given large responses in two choice taste or odour tests. Depressions in intake occurred with five added contaminants and, although internal metabolic effects were involved in three of these, those for glycine and malonic acid appear to be sensory responses. Two odours increased, and three depressed, intake. Glycine produced a similar reduction in intake both in the food and as an odour. Coumarin gave a much smaller depression as an odour contaminant than as an odour + taste contaminant from which it is apparent that the physiological effect of coumarin on digestion in the rumen is the main cause of the observed intake depression.

That "palatability" *per se* will influence the voluntary intake of foods has also been shown by Greenhalgh and Reid (1967) and by Welton and Baumgardt (1970). There is, however, little overall correlation between "palatability" ranking of a number of food sources and their ranking for food intake level when they are the sole source of food (Arnold, 1964b; Reid and Jung, 1965).

B. OTHER CHEMORECEPTORS

Not all the chemoreceptors censoring the chemical input into the ruminant are in the tongue and nose. Others are known to be in the oesophagus and other parts of the alimentary tract. Those in the oesophagus may give immediate feedback to the brain, or their signals may be integrated with those from the tongue. Feedbacks, either from chemoreceptors in the alimentary tract or from levels of blood metabolites, telling of the need of the animal for a particular nutrient or of the ill effects of ingestion of a particular plant may operate more slowly. A ruminant grazing 8–12 hours a day could produce feedbacks that amend its food selection during a day's grazing. There is, however, no evidence on this effect.

C. NUTRITIONAL WISDOM

Just how wisely the ruminant is able to select its food is often debated. Few, if any, critical experiments have been made from which this question can be answered. Whilst the persistence of a species must indicate that, on balance, the sensory responses have been developed to give adequate nutrition, the ruminant, like all other animals, is hedyphagic and may often select plants not for their nutritional advantages but for their flavour. Possibly, and there is no evidence other than for rats, the sensory responses of wild ruminants are more precisely fitted to the nutritional needs of the species than is so with domestic species. Domestic species are undoubtedly subjected by man in their feeding to sensory stimuli outside their inherent sensitivity ranges, which could explain why sheep, but not native kangaroos, in Western Australia preferentially eat plants of the *Gastrolobium* and *Oxylobium* species, whose monofluoracetate content is lethal. In the wild environment only animals which reject such plants, through some sensory response such as smell or taste, would survive and thus, in this case, "nutritional wisdom" would be a chance element in selection for survival.

It may thus be unreasonable to ask "do ruminants preferentially select plants to meet specific nutrient needs"?, or "do they preferentially avoid poisonous plants"? There is evidence on both sides but lack of really relevant data makes it impractical to discuss this issue further.

VII. PLANT CHEMICAL COMPOSITION EFFECTS ON FOOD SELECTION

It has been said that ruminants respond to stimuli from compounds present in plants. What is known of the sensory responses of ruminants to various pure chemicals has been described. There remains the major task of using this information to assess what has been studied in the field, and to project what still needs to be done. There is no doubt that failure to understand how and why ruminants select their food has led to enormous waste of effort in seeking for "palatability factors" in pasture plant species.

A. PROXIMATE COMPOSITION

There have been numerous attempts to relate preferences to the proximate composition of plants. Unfortunately it is not possible for the animal to recognize such things as nitrogen, "crude fibre", "energy" silica or "ash" as these fractions do not exist in this form at molecular level in the plant. Where correlations are found between any of these characteristics and preferences, they must relate to specific compounds or some physical property of the plant. In the case of ash, it could be related to the content of sodium or potassium salts; with fibre, the ease of harvesting could be the significant factor (Evans, 1964). Animals can not select a food purely because it is high in "energy" or produces the best "weight gain".

In analysing reports in the literature it is important to realize that lack of correlation between the content of a single chemical compound with preference rating is not proof of the unimportance of that compound in determining preferences. Because of the multidimensional nature of the process, all that is indicated is that it is not of paramount importance. Conversely, correlation between a single compound and preference is not proof that it is the main component influencing preference unless all other components have been accounted for. Even in closely related varieties, changes in accumulation on one biosynthetic pathway give no indication that all other pathways are unchanged.

In considering the relationship between the chemicals influencing taste and the chemical composition of the plant, little attention seems to have been given to the chemical form of the components in the plant tissue or to what proportion of these make contact with the senses. Most results are based on the total concentration of a component, or group of components, frequently determined on oven-dried material where the chemical form of the components will be quite different from that in the living tissue. The influences of plant and saliva enzymes are not considered and are largely unknown, a major problem being to account for the differences in the amount of chewing by individual animals at the time of ingestion. Both type and amount of enzymes and amount of chewing could well determine the chemical forms present at the taste-buds, and thus could be the deciding factor in acceptance or rejection.

The following eight groups of compounds have received attention to date in animal feeding experiments.

B. SUGARS

For the reasons mentioned, interpretation of ruminant responses to plant carbohydrates might best be described as confused. It is highly unlikely that an animal could give an integrated response to "soluble carbohydrates". What this fraction includes depends not only on the particular carbohydrates present in the plant under consideration, but also on the methods by which they are extracted and determined. Extraction methods alone show considerable variation; most techniques while adequate for energy considerations, are unsatisfactory for sweetness measurements. It may seem more likely that the amounts of hexoses and pentoses could be related to preferences on the basis of sweetness, but then these may be present as a wide range of stereoisomers of mono-, di-, tri- or polysaccharides, glycosides or other derivatives, each of which gives a different sweetness response. Even pure glucose in solution occurs in seven interconverting forms each with its own specific stereochemistry.

The type of sugar can influence ruminant preferences. Calves are known to take much more sucrose solutions in their TFI than other sugars at a given molarity (Kare and Ficken, 1963), and from Fig. 1 it seems that sheep prefer glucose to sucrose at high concentrations.

Reid et al. (1966), and Reid et al. (1967) noted significant positive correlations between preference rating and "soluble carbohydrate" content (ethanol soluble) with sheep on cocksfoot, but not with sheep on tall fescue (Reid and Jung, 1965).

Bland and Dent (1964) compared the preferences of cattle amongst fourteen strains of Dactylis glomerata at two locations at several times of the year. For early spring growth at one site, preferences were most closely correlated with per cent "total sugars", with hexoses and with fructoses but not with sucrose content. At all other observations "total sugar" content and preferences were not related.

With grass sorghum, Gangstad (1964) obtained a similar result in that preference rating of cattle was correlated with per cent "total sugar", but not with per cent "soluble carbohydrate" (in this case mainly starch). He concluded that cattle like high levels of "total sugar" and low levels of starch.

Marten and Donker (1964) showed that cattle would eat unacceptable "dung patch" herbage when it was sprayed with sucrose or molasses. A similar effect was demonstrated by Hironaka and Bailey (1968) who increased salt intake by adding "sugar", as quoted previously. Thus, for cattle, objectionable stimuli can be suppressed or overridden by the presence of sugar.

Since there are no reports of decreased preferences with increased sugar content, it seems likely that sugar content is important in determining ruminant preferences; but then sweetness is not confined to sugars, and certain sugar forms can give responses other than sweetness for non-ruminants.

C. ORGANIC ACIDS

The only published work (Jones and Barnes, 1967) showed a close positive correlation between sheep and cattle preferences and the citric plus shikimic acid contents of eight temperate grasses. This was repeatable at two different seasons of the year. Sheep, as shown in a previous section, have marked preference curves given a multi-choice amongst solutions of these acids, but the most acceptable molar concentrations differ quite markedly (aconitic 0·29, quinic 26·0, citric 23·8, malonic 0·48, fumaric 4·31, succinic 42·3, malic 3·73 mM). Because of this, differences in concentrations of these acids might well influence preferences. However, the present authors (unpublished data) have not been able to show any relationship between preference ranking and contents of any organic acids amongst twenty lines of Phalaris tuberosa and P. arundinacea, even for those with a low alkaloid content.

D. TANNINS

Tannins are known to depress digestion in the rumen and to be closely related to preference rating amongst strains of sericea (*Lespedeza cuneata* Don). Wilkins *et al.* (1953) demonstrated a 70% drop in voluntary intake of sericea herbage by cattle when tannin content increased from 4·8 to 12·0%. Donelly (1954) showed that steer preferences amongst 1206 lines of the species were related to tannin content and to stem fineness. Burns (1966) found that in sericea the astringent tannins were leucoanthocyanins (leucocyanidin and leucodelphinidin) and catechins. The ratio of astringent to total tannins could not be altered by any environmental factor affecting plant growth, nor does it vary genetically.

E. COUMARINS

To humans, coumarin has a sweet smell and bitter taste. The smell is objectionable to sheep and even a low level in the diet drastically reduces food intake (see above). Ashton and Jones (1959) sprayed ryegrass-white clover herbage to give it 0·5 to 1·0% coumarin. They observed that sheep, whilst initially (2–3 hours) avoiding the sprayed herbage, eventually consumed it. The difficulty with such a study is that a single extra component is added to the complex of taste and odour of the herbage. Since ryegrass-white clover herbage is usually very acceptable to sheep, it may be interpreted that the level of coumarin was insufficient to alter markedly the acceptability of this herbage. This could have been due to adjustment of the smell threshold, as has been described earlier. Alternatively the coumarin concentration on the herbage may have been reduced by evaporation or absorption and metabolism by the plant material.

A precursor of the coumarins, a compound of *o*-hydroxycinnamic acid, which occurs in *Melilotis*, is said to lower the acceptability of the herbage of this species, but there is no evidence of this. There is, however, evidence of physiological effects on sheep and cattle by the conversion of this acid to dicoumarol.

F. ISOFLAVONES

In the genus *Trifolium*, many species contain isoflavones in quite large quantities. In *T. subterraneum*, an important pasture plant in Australia, up to 10% of the plant dry weight may be isoflavones in the free form and as isoflavone glycosides (Hill and Arnold, unpublished). The authors have recently compared preferences of sheep amongst nine lines of *T. subterraneum* over two growing seasons. Neither total isoflavones nor amounts of any individual isoflavones were related to preferences. Also, Francis and Beck (private communication) have found that a high isoflavone strain and a derived mutant lacking isoflavone both have the same acceptability to sheep. Thus, although these isoflavones can cause serious reproductive disorders in sheep, sheep show no discrimination against strains of clover containing them.

G. ALKALOIDS

A number of different alkaloids are known to reduce acceptability of strains of several food plants. In most cases, these alkaloids also have physiological effects on ruminants, either through impairing digestion in the rumen, or directly on the mammalian system. Probably the discrimination in selection of alkaloids represents the clearest case of preferences being physiologically wise, although this is not always so.

1. Quinolizidine Alkaloids

These alkaloids are widespread in the leguminous genera: *Lupinus, Baptisia, Cytisus* and *Genista*.

In the genus *Lupinus*, some thirty different but related alkaloids occur, both sweet (alkaloid-free) and bitter (alkaloid-containing) strains occurring within the same species. The presence of alkaloids is genetically controlled. The main alkaloid in *L. angustifolius* is lupanine, which can be as high as 2·5% of the herbage dry matter (Gordon and Henderson, 1951). Sheep avoid grazing alkaloid lines of this species, if at all possible, but readily graze alkaloid-free varieties. According to Gladstones (1970), sweet and bitter strains of a species differ only by a single gene so that other chemical components may be similar. If this is so, then alkaloids may well be the only factor influencing acceptability. Further work is required to clarify this situation.

2. Pyrrolizidine Alkaloids

These are found in several important herbage species, particularly *Lolium* spp. and *Festuca arundinacea*. The main alkaloid is perloline, but festucine (Yates and Tookey, 1965) may also be present. As yet, there is no evidence that these affect acceptability although perloline depresses digestion in the rumen (Bush *et al.*, 1970).

3. Indolealkylamine Alkaloids

A great deal of work has been done recently on these alkaloids in *Phalaris tuberosa* and *P. arundinacea* because of their known physiological effects and their suspected effects on acceptability of certain strains of the species.

Both Roe and Mottershead (1962) and Kennedy and Holgate (1971) showed that the low acceptability of certain strains of *P. arundinacea* was due to a water-soluble chemical compound. Brown (1961) found that preferences amongst strains of this species were correlated with the content of water-soluble phenolic constituents—a large group of compounds. O'Donovan *et al.* (1967) found that acceptable and unacceptable lines did not differ in water-soluble carbo-hydrates, but the unacceptable lines had more water-soluble nitrogen in their total nitrogen, which could indicate alkaloids or their precursors.

P. tuberosa and *P. arundinacea* differ in alkaloid composition. *P. tuberosa* contains predominately *NN*-dimethyltryptamine (DMT), 5-methoxy-*NN*-dimethyltryptamine and bufotenine (Culvenor, Bon and Smith, 1964) whilst *P. arundinacea* contains mainly gramine and hordenine, with smaller amounts of other alkaloids.

Barnes *et al.* (1969), in a brief report, suggest that in *P. arundinacea* palatable lines have up to 0·1 % gramine and no DMT, whereas unpalatable lines have no gramine but up to 0·02% DMT.

A *P. tuberosa* strain low in all alkaloids was sprayed with gramine to give four levels of gramine on a dry matter basis, and fed to sheep (Arnold unpublished). From Table IV it is clear that gramine at low levels stimulates intake, but at higher levels depresses it. The effect with 1·0% gramine agrees with that given in Table III. However, when levels of 0, 0·01 and 0·20% gramine were sprayed on herbage and sheep were given free choice of the three levels, they consumed equal amounts of the three. This illustrates that preference, or lack of it, amongst a number of choices is often a poor guide to what food intake will be for the same feeds when there is no choice, as mentioned earlier.

TABLE IV

Effect of gramine on the dry matter intake of
Phalaris tuberosa by sheep

Gramine as % of dry matter	Mean daily dry matter intake (g)
0·00	1144 b
0·01	1295 a
0·10	1009 c
1·00	668 d

In some current work, the authors are comparing 20 strains of *P. tuberosa*, *P. arundinacea* and some interspecific hybrids for acceptability. A broad spectrum of chemical constituents is being examined in relation to the measured acceptability, including the individual organic acids from dicarboxylic to phenolic, free sugar and aminonitrogen content and total alkaloids. For seven harvests over two years, the largest amount of variation in acceptability is accounted for in multiple regressions by total alkaloids with individual alkaloid composition appearing to be of little significance. The evidence to date points to the indolealkylamines dominating preferences in both species.

H. ESSENTIAL OILS

Many species that are important food plants for browsing ruminants contain essential oils. In California, Oh and his colleagues have been studying these in relation to deer and sheep preferences. They have found (Oh *et al.*, 1967) that the monoterpene hydrocarbons have little effect, the monoterpene esters a mild effect, but that the monoterpene alcohols strongly inhibit rumen activity. Whether the effects on the rumen are related to palatability differences is not yet established.

I. FATS

There are several reports of situations where preferences of ruminants are correlated with the ether extract content of plants on offer. The most striking is that of Louw *et al.* (1967) who found that sheep preferences amongst species of bush in the Karoo of South Africa were related to their ether extract content, decreasing with increasing fat content. Unfortunately, there has been no attempt to identify which fats, if any, in their complex extracts influence acceptability.

VIII. The Potential Taste and Odour Constituents of Plants

In the previous section, consideration was given only to those compounds or groups of compounds that had been studied in relation to ruminant food preferences. Undoubtedly there are many more of importance. The list in Table V summarizes these data and indicates some chemicals in herbage plants that should be considered further. Where there is information, maximum levels found and sheep taste and odour responses are given, together with the approximate concentration for taste rejection. A limitation of this table is that the data shown for acids are for acids dissolved in water. Although the form of many of these acids in plants is unknown, it may be assumed that many will be as esters. As Table VI shows, at equal molarity, sheep prefer solutions of the sodium salts of the acids to the acids themselves.

A. ENVIRONMENTAL INFLUENCES

The variation in concentration of each constituent listed is great and determined not only by genetic factors but also by environmental factors. For example, *NN*-dimethyltryptamine levels in *Phalaris tuberosa* are highest under high temperature, low light intensity and high nitrogen status (Moore *et al.*, 1967). Other alkaloids follow a similar trend with nitrogen fertilizer (e.g. perloline—in tall fescue—Gentry *et al.*, 1969; and in ryegrass—Bennett, 1963). Likewise, organic acids in ryegrass increase with level of nitrogen fertilizer

TABLE V

Plant chemical constituents and their taste and odour to sheep

Constituents	Molarity of taste rejection* (mM)	Odour response	Maximum levels reported as % of plant dry weight	Approximate mM concentration on whole plant basis	Plant species	Reference
Volatiles						
Ethanol						
Propanol						
Acetone	0				Medicago sativa	
2-Methylpropanol					Trifolium pratense	Woods and Aurand (1963);
Butanol		−ve			T. repens	Morgan and Pereira
Butanone					Phleum pratense	(1963)
3-Methylbutanol					Bromus	
Pentanal					Lolium perenne	
n-Hexanal						
trans-2-Hexenal		+ve				
cis-3-Hexenal						
n-Hexanol		−ve				
trans-2-Hexenol		0				
cis-3-Hexenol		−ve	0·00007	—	Trifolium repens	
1-Octen-3-ol			0·0002	—	T. hybridum	Honkanen et al. (1969)
1-Octen-3-one			0·000015	—	T. hybridum	
Saccharides						
Xylose			0·03	4	Trifolium repens	
Glucose			0·08	9	T. repens	Bailey (1958)
Fructose			0·09	10	T. repens	

	Value	Count	+ve/−ve	Species	Reference
Sucrose	5·8			*Dactylis glomerata*	Waite (1970)
Raffinose	34			*Medicago sativa*	Hirst and Ramstad (1957)
Stachyose				*Lolium perenne*	Laidlaw and Reid (1952)
Amino Acids					
Glycine	0·15	4		*Lupinus albus*	Rabideau (1952)
Alanine	1·37	31		*L. albus*	Fauconneau (1960)
Valine	0·60	10		*Bouteloua curtipendula*	Rabideau (1952)
Leucine	0·80	12		*B. curtipendula*	Fauconneau (1960)
Serine	0·14	3		*Medicago sativa*	Edel'Shtein (1968)
Threonine	0·39	7		*L. albus*	
Phenylalanine	0·33	4		*L. albus*	Rabideau (1952)
Tyrosine	0·20	2		*L. albus*	
Aspartic acid	0·80	12		*B. curtipendula*	Fauconneau (1960)
Glutamic acid	0·80	11		*Sorghastrum nutans*	
Asparagine	0·42	6		*M. sativa*	Rabideau (1952)
Glutamine	1·87	26		*Dactylis glomerata*	
Lysine	0·70	10		*B. curtipendula*	Fauconneau (1960)
Arginine	0·70	8		*B. curtipendula*	
Tryptophan	0·18	2		*L. albus*	Edel'Shtein (1968)
Proline	0·14	2		*M. sativa*	
γ-Aminobutyric acid	0·34	7		*M. sativa*	Fauconneau (1960)
Polybasic aliphatic Acids					
Malonic	96		+ve	*Medicago sativa*	Fauconneau (1958)
Succinic	85			*Lolium perenne*	Jones and Barnes (1967)
Fumaric	86			*Dactylis glomerata*	Fauconneau (1960)
Malic	149		−ve	*D. glomerata*	Stout et al. (1967)
Trans-Aconitic	29			*Phalaris tuberosa*	Hirst and Ramstad (1957)
Citric	238			*L. perenne*	

TABLE V—*continued*

Constituents	Molarity of taste rejection* (mM)	Odour response	Maximum levels reported as % of plant dry weight	Approximate mM concentration on whole plant basis	Plant species	Reference
Polyhydroxy Acids						
Quinic	52	+ve	1·5	16	*Lolium perenne*	⎫ Jones and Barnes (1967)
Shikimic			0·5	6	*Dactylis glomerata*	⎬
2-Hydroxycinnamic			6·2	75	*Melilotus alba*	Whited *et al.* (1966)
Coumarin		-ve	4·4	60	*M. alba*	Smith (1964)
Chlorogenic			1·04	6	*L. perenne*	Martin (1970)
"Tannic"	(0·1%)	-ve	15·2		*Lespedeza cuneata*	Burns (1966)
Isoflavones						
Formononetin			3·1	23	*Trifolium subterraneum*	G. W. Arnold and J. L. Hill
Daidzein			0·6	5	*T. subterraneum*	(unpublished)
Biochanin A			5·7	40	*T. subterraneum*	
Genistein			6·2	46	*T. subterraneum*	
Alkaloids						
Gramine			0·3	3·4	*Phalaris arundinacea*	Culvenor *et al.* (1964)
Hordenine			0·009	0·1	*P. arundinacea*	Audette *et al.* (1970)
Bufotenine			0·008	0·08	*P. tuberosa*	Oram (unpublished)
NN-Dimethyltryptamine (DMT)			0·087	0·92	*P. tuberosa*	Oram and Williams (1967)
5-MethoxyDMT			0·120	1·1	*P. tuberosa*	⎫ Moore *et al.* (1967)
5-HydroxyDMT			0·009	0·09	*P. tuberosa*	⎬

* As determined in solution multi-choice experiments.

TABLE VI

Preferences of sheep between acid and salt solutions

Solutions (mM concn)		% of total fluid intake as acid solution
Na Citrate:Citric acid		
23·8	23·8	36·9
Na Succinate:Succinic acid		
42·3	42·3	36·7
Na Glutamate:Glutamic acid		
34·0	34·0	27·7

(Waite, 1970), although the proportions of the various acids in plants will change with different nutritional deficiencies (Cummings' and Teal, 1965). In sweet clover the o-hydroxycinnamic acid content generally increases with increased light intensity and increased photoperiod (Whited et al., 1966).

B. UNKNOWN RESPONSES

Two striking omissions from the field studies of plant chemicals and food preferences have been the free amino acids and most aliphatic volatile compounds. This may be because these substances are expensive when bought in the quantities needed for such studies. It is of interest that, to humans, the grassy aroma in "tainted" milk is due to trans-2-hexenol.

Some forty volatile flavour substances have been identified from Trifolium pratense L., T. hybridum L., and T. repens L. (Honkanen et al., 1969) but until more specific odour response data are available for ruminants, no further comment can be made.

C. THEORETICAL STUDIES

There is a long way to go if we want to classify plant chemicals in ruminant taste responses in terms of criteria such as sweetness, saltiness, bitterness and sourness. However, experimental means may be available.

Tapper and Halpern (1968) have developed a technique with rats in which they showed that rats classified solutions of sodium saccharine, DL-alanine and glycine similarly, but differently from solutions of glucose. A more basic physical-chemistry approach was taken by Shellenberger et al. (1969) and Shellenberger and Acree (1969): they propose that the sweet taste of the individual anomeric forms of D and L sugar and amino acid enantiomers is due to a bifunctional group consisting of an acidic (AH) and a basic (B) moiety

5

with an A–H proton to B distance of about 3 Å. This fits into the taste bud receptor as shown below:

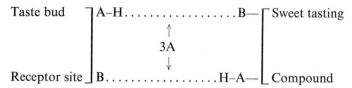

On this basis they can predict whether a compound will be sweet or not. Unless receptor mechanisms differ between mammalian species, then this basic principle should apply to ruminants. Its use will depend on detailed knowledge of the stereochemistry and molecular forms of compounds producing responses at the tastebuds.

Horowitz and Gentili (1969) have studied taste and structure in phenolic glycosides and show how, from alterations at selected sites in flavone neohesperidosides, taste varies from bitter to bitter-sweet to sweet to tasteless.

The development of such approaches should result in logical predictions of the taste of chemicals for use in studying food preferences in ruminants. Naturally, calibrations for each animal species will be essential.

IX. CONCLUSIONS

In 1952, Tribe was of the opinion that "to relate the chemical composition of foods to their degree of palatability is difficult and perhaps at the moment impossible. We may assume that chemical changes cause corresponding changes in the taste and odour of a plant, but a quantitative estimation of such changes or an assessment of their importance in palatability has never been made".

Nineteen years later we come to a very similar conclusion. Perhaps the only hopeful thought is that the basis for a more thorough understanding of how chemical factors determine selection of food plants by ruminants is being established. Knowledge of the physiology of taste, and to a less extent of smell, in ruminants has advanced. It has been shown that both the smell and taste of plants influence their selection. There is a growing realization that ruminants cannot respond to groupings of chemical constituents that are based on the arbitrary divisions convenient to analysis. They can only respond to molecular concentrations of individual compounds in the form in which they occur in plants and only to those that the animals' chemical receptors accept. Failure to recognize this fact in the past has meant that few of the papers in which relationships between food preferences and plant chemical composition are discussed give relevant information on cause and effect.

It is possible to deduce, in the cases of certain alkaloids and tannins, that preferences shown between strains of a plant species are almost certainly due to unfavourable responses to these compounds. However, it is not possible to attribute these to odour or taste, or both. Nor is it known how the responses are modified by the presence of other chemicals. In future studies such information is needed. Knowledge of metabolic pathways is essential if we want to know what chemical forms are likely to be present, how pathways differ for different plant species or strains of species, and how these pathways can be modified, e.g. by genetic selection or nutritional regimes, to obtain acceptability of the plant by the grazing animal.

New methods need to be developed to elucidiate the role (if any) of "nutritional wisdom" as it applies to ruminants both in pen and in grazing situations. The significance of feedback mechanisms in controlling selection and rejection thresholds and their consequent effect on selection and intake is poorly understood.

This type of data, together with more data on the chemical structures that give odour and taste responses in ruminants, should allow predictive models of food preferences to be built. Once this is achieved it will be possible to analyse plants for those chemicals known to influence preferences and not, as at present, for chemicals thought to be influential.

In the meantime, it should be possible to develop a more logical approach to identifying the cause of low acceptability of particular plant cultivars. This, it is suggested, should involve testing the animal response to carefully prepared extracts from known unacceptable and acceptable strains. When an extract from the unacceptable strain, but not one from the acceptable strain, is shown to reduce acceptability, its significant chemical components should be isolated and identified, and each constituent tested at various concentrations to measure its effects. Once identified, the factors influencing the concentration of the significant compounds can be determined and their genetic control explored.

REFERENCES

Arnold, G. W. (1964a). In "Grazing in Terrestrial and Marine Environments" (D. J. Crisp, ed.), pp. 133. Blackwell Scientific Publications, Oxford.
Arnold, G. W. (1964b). Proc. Aust. Soc. Anim. Prod. 5, 258.
Arnold, G. W. (1966a). Aust. J. agric. Res. 17, 531.
Arnold, G. W. (1966b). Aust. J. agric. Res. 17, 521.
Arnold, G. W. (1970). In "Physiology of Digestion and Metabolism in the Ruminant" (A. T. Phillipson, ed.), pp. 264–276.
Ashton, W. M. and Jones, E. (1959). J. Brit. Grassl. Soc. 14, 47.
Audette, R. C. S., Vizayanagar, H. M., Bolam, J. and Clark, K. W. (1970). Can. J. Chem. 149.
Bailey, R. W. (1958). J. Sci. Fd Agric. 9, 748.
Baldwin, B. A., Bell, F. R. and Kitchell, R. L. (1968). Proc. Physiol. Soc. 14P.
Barnes, R. F., Martin, G. C. and Simons, A. B. (1969). Agron. Abstr. p. 57.
Baryshnikov, I. A. and Kokorina, E. P. (1964). Dairy Sci. Abstr. 26, 97.
Bell, F. R. (1959). J. agric. Sci. Camb. 52, 125.
Bell, F. R. and Williams, H. L. (1959). Nature, Lond. 183, 345.
Bennett, W. D. (1963). N.Z. agric. Res. 6, 310.
Bernard, R. A. and Kare, M. R. (1961). J. anim. Sci. 20, 965.

Bland, B. F. and Dent, J. W. (1964). *J. Br. Grassl. Soc.* **19**, 306.
Brown, J. A. M. (1961). *Diss. Abstr.* **22**, 373.
Burns, R. E. (1966). Bu NS1964 *Georgia Agric. Exp. Stn.*
Bush, L. P., Streeter, C. and Buckner, R. C. (1970). *Crop Sci.* **10**, 108.
Crawford, J. C. (1970). M.S. Thesis, Oregon State University, Corvallis, Oregon.
Culvenor, G. C. J., Bon, R. D. and Smith, L. W. (1964). *Aust. J. Chem.* **17**, 1301.
Cummings, G. A. and Teal, M. R. (1965). *Agron. J.* **57**, 127.
Donelly, E. D. (1954). *Agron. J.* **46**, 96.
Edel'Shtein, M. M. (1968). *Vestn. Sel'skokhoz. Nauki (Moscow)* **13** (4), 26.
Evans, P. S. (1964). *N.Z. J. Agric. Res.* **7**, 508.
Fauconneau, G. (1958). *Ann. Agron. Suppl.* pp. 1–13.
Fauconneau, G. (1960). *Proc. 8th Int. Grassl. Congr.* pp. 617–20.
Gangstad, E. O. (1964). *Crop Sci.* **4**, 269.
Gentry, C. E., Chapman, R. A. and Henson, L. (1969). *Agron. J.* **61**, 313.
Gladstones, J. S. (1970). *Field Crop Abstr.* **23**, 123.
Goatcher, W. D. and Church, D. C. (1970a). *J. anim. Sci.* **30**, 777.
Goatcher, W. D. and Church, D. C. (1970b). *J. anim. Sci.* **30**, 784.
Goatcher, W. D. and Church, D. C. (1970c). *J. anim. Sci.* **31**, 364.
Goatcher, W. D. and Church, D. C. (1970d). *J. anim. Sci.* **31**, 373.
Goatcher, W. D., Church, D. C. and Crawford, J. (1970). *Feedstuffs* **42** (47), 16.
Gordon, J. G. (1970). *Proc. Nutr. Soc.* **29**, 325.
Gordon, W. C. and Henderson, J. H. M. (1951). *J. agric. Sci. Camb.* **41**, 141.
Greenhalgh, J. F. D. and Reid, G. W. (1967). *Nature, Lond.* **214**, 744.
Heady, H. F. (1964). *J. Rge. Mgmt.* **17**, 76.
Hironaka, R. and Bailey, C. B. (1968). *Can. J. Anim. Sci.* **49**, 455.
Hirst, E. L. and Ramstad, S. (1957). *J. Sci. Fd. Agric.* **8**, 727.
Honkanen, E., Moisio, T. and Karvonen, P. (1969). *Suom. Kemistilehti* **42**, 448.
Horowitz, R. M. and Gentilii, B. (1969). *J. agric. Fd. Chem.* **17**, 696.
Jones, E. C. and Barnes, R. J. (1967). *J. Sci. Fd. Agric.* **8**, 321.
Kare, M. R. (1969). *J. agric. Fd. Chem.* **17**, 677.
Kare, M. R. and Ficken, M. S. (1963). *In* "Olfaction and Taste". (Y. Zotterman, ed.), pp. 285–97.
Kennedy, G. S. and Holgate, M. D. (1971). *Aust. J. exp. Agric. Anim. Husb.* (in press).
Kudryavtsev, A. A. (1962). *Proc. 16th Int. Dairy Congr.* pp. 565–72.
Laidlaw, R. A. and Reid, S. G. (1952). *J. Sci. Fd. Agric.* **3**, 19.
Louw, G. N., Steenkamp, C. W. P. and Steenkamp, E. L. (1967). *S. Afr. Tydsk. Landbouwet.* **10**, 867.
Marten, G. C. and Donker, J. D. (1964). *J. Dairy Sci.* **47**, 773.
Martin, A. K. (1970). *Br. J. Nutr.* **24**, 943.
Moore, R. M., Williams, J. D. and Chia, Joyce (1967). *Aust. J. Biol. Sci.* **20**, 1131.
Morgan, M. E. and Pereira, R. L. (1963). *J. Dairy Sci.* **46**, 1420.
O'Donovan, P. D., Barnes, R. F. and Plumlee, M. P. (1967). *Agron. J.* **59**, 478.
Oh, H. K., Sakai, T., Jones, M. B. and Langhurst, W. M. (1967). *Appl. Microbiol.* **15**, 777.
Oram, R. N. and Williams, J. D. (1967). *Nature, Lond.* **213**, 946.
Rabideau, G. S. (1952). *Bot. Gen.* **113**, 475.
Reid, R. L. and Jung, F. A. (1965). *J. anim. Sci.* **24**, 615.
Reid, R. L., Jung, G. A. and Murray, S. J. (1966). *J. anim. Sci.* **25**, 636.
Reid, R. L., Jung, G. A. and Kinsey, C. M. (1967). *Agron. J.* **59**, 519.
Roe, R. and Mottershead, B. E. (1962). *Nature, Lond.* **193**, 255.
Shellenberger, R. S., Acree, T. E. and Lee, C. Y. (1969). *Nature, Lond.* **221**, 555.
Shellenberger, R. S. and Acree, T. E. (1969). *J. agric. Fd. Chem.* **17**, 701.
Smith, W. K. (1964). *Crop Sci.* **4**, 666.

Stout, P. R., Brownwell, J. and Burau, R. G. (1967). *Agron. J.* **59**, 21.
Stubbs, O. J. and Kare, M. R. (1958). *J. anim. Sci.* **17**, 1162.
Tapper, D. N. and Halpern, B. P. (1968). *Science, N.Y.* **161**, 708.
Tribe, D. E. (1949). *J. agric. Sci. Camb.* **39**, 309.
Tribe, D. E. (1952). *Proc. 6th Int. Grassl. Congr.* pp. 1265.
Troelsen, J. E., Muhr, P. I., Lodge, R. W. and Kitcher, M. R. (1968). *Can. J. anim. Sci.* **48**, 373.
Waite, R. (1970). *J. agric. Sci. Camb.* **74**, 457.
Welton, R. F. and Baumgardt, B. R. (1970). *J. Dairy Sci.* **53**(12), 1771.
Whited, D. A., Gorz, H. J. and Hoskins, F. A. (1966). *Crop. Sci.* **6**, 73
Wilkins, H. L., Bates, R. P., Hewson, P. R., Lindahl, I. L. and Davis, R. E. (1953). *Agron. J.* **45**, 335.
Woods, A. E. and Aurand, L. W. (1963). *J. Dairy Sci.* **46**, 656.
Yates, S. G. and Tookey, H. L. (1965). *Aust. J. Chem.* **18**, 53.

CHAPTER 7

Cyanogenic Glycosides and Their Function

DAVID A. JONES

Department of Genetics, The University, Birmingham, England

I. INTRODUCTION

For many years, Fraenkel (1959) has been concerned about the metabolic function of secondary plant substances. He groups together glycosides, saponins, tannins, alkaloids, essential oils, and others and notes that, whereas their occurrence is generally sporadic, individual substances may, however, be specific to families, subfamilies, genera and even to species and subspecies. He comments that their function in metabolism has never been satisfactorily explained and he argues that, because of their sporadic occurrence, it is unlikely that they play a basic rôle in the physiology of plants. He suggests that the food specificity of insects is based solely upon the presence or absence of these odd compounds, whereby they serve as repellents to insects and other animals in general, and as attractants to those few in particular which feed on each plant species. He does not imply that these plant substances are necessarily used in tissue synthesis by the animals. Ehrlich and Raven (1965) focused their discussion on the relations between butterflies and their larval food–plants, and they have put forward the following scheme: "We suggest that some of these (*secondary plant*) compounds may have been present in early angiosperms and afforded them an unusual degree of protection from the phytophagous organisms of the time, relative to other contemporary plant groups. Behind such a biochemical shield the angiosperms may have developed and become structurally diverse. . . . In turn, the fantastic diversification of modern insects has developed in large measure as the result of a stepwise pattern of coevolutionary stages superimposed upon the changing pattern of angiosperm variation." The problems of interactions between species, at the chemical level, has been taken even further by Whittaker and Feeny (1971) and by Levin (1971).

Stahl (1888) appears to have been the first person to suggest that plants possess chemical defense mechanisms, as well as the morphological and mechanical ones with which biologists were familiar, and Fraenkel uses this author as his source of inspiration. Even as recently as 1969, Fraenkel wrote: "I know for certain that the army of biochemists and organic chemists who are engaged in unravelling the structure and metabolic function of these compounds, is totally unaware of, and uninterested in, the reasons for their existence." This article is an attempt to redress this balance, but at the outset it is clear that Fraenkel has overlooked the work of Conn and his co-workers. Six years ago Abrol and Conn (1966) briefly considered "a possible metabolic role for the cyanogenic glucosides" and in their article on the "Biosynthesis of Cyanogenic Glycosides and other simple Nitrogen Compounds", Conn and Butler (1969) elaborate the discussion of the significance of cyanide metabolism in plants.

In her review of cyanogenesis in plants, Muriel Robinson (1930) considered three possible functions for cyanogenic glycosides: excretions, metabolism of nitrogen and defence. These are the limits of the terms of reference for this article because no new functions have been suggested since that time.

II. Cyanogenesis and Cyanogenic Glycosides

As can be seen from the early history of the study of the cyanogenesis outlined in Table I, the ability of some plants to produce hydrogen cyanide has been recognized for many years. Paris (1963) comments that although many plants yield HCN when treated with enzyme, in many cases it is in no way certain that cyanogenic glycosides are the source of the HCN. This is a sensible approach to take and is fully justified by the recent clarification that the cyanogenic substances in *Cordia verbenacea* DC. and *Ungnadia speciosa* Endl. (Sapindaceae) are lipids (see e.g. Seigler *et al.*, 1971). Gibbs (1963) presents a long list of cyanogenic glycosides, pointing out their occurrence and their uses in chemical plant taxonomy. He concludes his article with an enormous list of references. Other accounts will be found in Wokes and Willimott (1951) and Kingsbury (1964).

Cyanogenic glycosides are widely distributed amongst 60 or so flowering plant families (Hegnauer, 1963), and are also found in several species of ferns (see for example Berti and Bottani, 1968), fungi (Guyot, 1916; Lebeau and

TABLE I

A brief history of cyanogenesis up to 1912.

Discovery	Date	Authority
Plants produce prussic acid	The dim and distant past	
Prussic acid from bitter almonds	1803	Schrader
Prussic acid from bark of *Prunus padus*	1812	Bergemann
Amygdalin isolated from bitter almonds	1830	Robiquet and Boutron Charlard
Hydrolysing enzyme (Emulsin) discovered	1837	Wöhler and Liebig
Survey of Rosaceae for cyanogenic species	1851	Wicke
Amygdalin in seeds of several species of Rosaceae	1870	Lehmann
HCN from fungi	1871	Lösecke
Cyanogenesis in *Arum maculatum* and *Linum usitatissimum*	1883–1884	Jorissen
Linamarin isolated from *Linum*	1891	Jorissen and Hairs
Large survey of flowering plants, cyanogenesis found in 20 natural orders	1896–1907	Treub
Isolation of dhurrin and lotusin, confirmation of discovery of linamarin	1901–1903	Dunstan and Henry
Cyanogenesis in *Lotus corniculatus*	1912	Armstrong, Armstrong and Horton
Cyanogenesis in *Trifolium repens*	1912	Mirande

For references other than Schrader (1803) and Lösecke (1871), see Robinson (1930).

Dickson, 1953) and bacteria (Clawson and Young, 1913; Patty, 1921). Some of these glycosides appear to be restricted to single genera; vicianin for example occurs in *Vicia* L. species only. Other substances are widespread, linamarin being found in the Linaceae (*Linum usitatissimum* L.), Euphorbiaceae (*Manihot utilissimus* Pohl., *Hevea brasiliensis* Müll.-Arg.), Compositae (*Dimorphotheca* Moench spp. and *Osteospermum jucundum* Nordlindh.), Leguminosae (*Lotus* L. spp., *Trifolium repens* L., *Phaseolus* L. spp. etc.) (Butler, 1965) and possibly in the Ranunculaceae (*Thalictrum aquilegifolium* L.) although Sharples and Stoker (1969) do not confirm this.

The distribution of cyanogenic glucosides over a wide range of families and the fact that their biosynthetic pathways are identical in three species from different families (Abrol and Conn, 1966) raises the old problem that the genetic systems involved could be very old, perhaps even as old as flowering plants. Sequence analyses of the amino-acids in the enzymes involved could quickly reveal whether this is an example of homology rather than analogy.

When we look closely at individual genera, we find that some species are cyanogenic while others are not. In the genus *Lotus*, *L. arabicus* L., *L. cornicu-*

5*

latus L., *L. creticus* L., *L. parviflorus* Desf. and *L. tenius*, Waldst. & Kit., are cyanogenic whereas *L. uliginosus* Schkuhr., *L. alpinus* (DC.) Schleicher ex Ramond (synonym *L. corniculatus* L. var. *alpinus* Ser.) and *L. japonicus* (Regel) Larsen are exclusively acyanogenic. A similar pattern occurs in the genus *Trifolium*. Furthermore, populations of *L. corniculatus* and *T. repens* contain both cyanogenic and acyanogenic plants and amongst cyanogenic plants there are quantitative differences between individuals. Thus there is variation within species, within genera, and within families with respect to cyanogenesis.

But cyanogenesis is not confined to plants. It is an aphorism that red colouration in Lepidoptera indicates that the beast is unpleasant to eat and Jones *et al.* (1962) have described the production of HCN at all stages in the life cycle of the moths *Zygaena filipendulae* L. and *Z. lonicerae* von Schev. Furthermore, it has been known for a long time that millipedes produce a defensive secretion and Eisner *et al.* (1963a,b) have now shown that the paired segmental glands in several species of the order Polydesmida produce cyanogenic substances. They also point out that the secretion acts as a powerful deterrent to attack by ants and other predators. There is little doubt, therefore, that cyanogenesis is a protective mechanism in these animals.

Up to now it has not been possible to detect free HCN outside intact cyanogenic plants other than fungi. The plant must be damaged physically by crushing or by freezing (Daday, 1962, 1965), or chemically by adding a mordant like toluene or chloroform. Plants which are phenotypically cyanogenic contain an enzyme (Wöhler and Liebeg 1837) or series of enzymes (Mao and Anderson, 1967; Butler *et al.*, 1965; Haisman and Knight, 1967; Stevens and Strobel, 1968) which are classed as "β-glycosidases" (EC. 3.2.1.21), in addition to the appropriate cyanogenic glycoside. It has been clear for many years that there are different kinds of β-glycosidase (Robinson, 1930). The enzyme from *Linum usitatissimum* which hydrolyses linamarin has little or no effect on amygdalin while "emulsin" from *Prunus amygdalus* Batsch is weakly active on linamarin. β-Glycosidases are widely distributed and have been recorded, for example, in *Escherichia coli* (Migula) Castelani et Chalmers (Schaefler, 1967), yeast (Duerksen and Halvorson, 1959) and mussels (Viebel *et al.*, 1963). There are several sources of information on methods of estimating β-glucosidases (also called β-D-glucoside glucohydrolases); the traditional techniques can be found in Barman (1969) while fluorimetric assays are described in Robinson *et al.* (1967).

There is some evidence about the distribution of the cyanogenic glycosides within the plant and on quantitative variation. In *Trifolium repens*, for example, the cyanogenic glucosides appear to be present only in leaflets and in petioles. The concentration seems to fluctuate diurnally (De Waal, 1942) and it rises during the season (Askew, 1933; Rogers and Frykolm, 1937) and also under dry conditions (Rogers and Frykolm, 1937) and is higher in induced tetraploids than in the clones from which they were produced (Hutton, 1957).

III. BIOSYNTHESIS OF CYANOGENIC GLYCOSIDES

As far as the biosynthesis and the hydrolysis of the cyanogenic glycosides are concerned, the reader is referred to Conn and Butler (1969) for a full and clear account. A common factor in all the pathways which have been examined

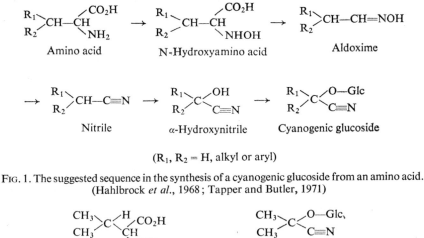

$$(R_1, R_2 = H, \text{ alkyl or aryl})$$

FIG. 1. The suggested sequence in the synthesis of a cyanogenic glucoside from an amino acid. (Hahlbrock *et al.*, 1968; Tapper and Butler, 1971)

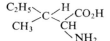

FIG. 2. The relationship between four cyanogenic glucosides and their amino acid precursors

up to now is that the sequence begins with an amino acid (Fig. 1). Four such pathways have now been inferred from ^{14}C labelling experiments and the cyanogenic glycosides and the amino acids concerned are listed in Fig. 2. The occurrence of individuals of *Trifolium repens* and *Lotus corniculatus* which apparently contain neither linamarin nor lotaustralin suggests that techniques similar to those using auxotrophic mutants in the elucidation of biosynthetic pathways in micro-organisms could be helpful for confirming this synthetic sequence.

Linamarin and lotaustralin appear to occur together more often than not (Butler, 1965). Indeed it is only in the rubber tree *Hevea braziliensis* that linamarin occurs in the apparent absence of lotaustralin. Is it possible to have cyanogenic forms of *T. repens*, let us say, which contain only one of these cyanogenic glucosides and not both or neither? Recent work by Hahlbrock and Conn (1970, 1971) suggests that this is unlikely, because they have evidence that both linamarin and lotaustralin are synthesized from amino acids using the same enzymes. Certainly the last step in the reaction outlined in Fig. 1 appears to be controlled by the same glucosyltransferase in both linamarin and lotaustralin synthesis (Hahlbrock and Conn, 1971).

IV. CYANIDE METABOLISM IN RELATION TO CYANOGENIC GLYCOSIDES

Studies on the biosynthesis of cyanogenic glycosides show that it is the α-carbon and nitrogen atoms of valine which become the nitrile group of linamarin (Butler and Butler, 1960); the remainder of the aglycone moiety derives from the rest of the amino acid. In addition, when $H^{14}CN$ is fed to plants, the CN^- groups becomes the amide group of asparagine (Blumenthal-Goldschmidt *et al.*, 1963). A crucial observation was made by Abrol and Conn (1966) working with *Lotus arabicus* and *L. tenuis*. They demonstrated a similarity in the labelling pattern of radioactive asparagine, isolated from plants fed uniformly labelled L-valine-U-^{14}C to that isolated from plants exposed to $H^{14}CN$. In consequence, they suggested that the nitrile moiety of linamarin can provide the $H^{14}CN$ for the biosynthesis of asparagine in an *intact plant*. Similarly in *Nandina domestica* Thunb. asparagine was synthesized from L-tyrosine via a cyanogenic glycoside, showing that it is not the aglycone as a whole which is involved. Blumenthal *et al.* (1968) have evidence, however, that the acyanogenic species *Lupinus angustifolius* L. does not use a cyanogenic glycoside as an intermediary in the synthesis of asparagine from isoleucine. Thus the possibility that the α-carbon and nitrogen of the first amino acid becomes the amide group of asparagine without the need for a cyanogenic intermediary is not ruled out. An inescapable conclusion to be drawn from the work of Abrol and Conn (1966) and Abrol *et al.* (1966) is that cyanogenic glycosides are metabolically active rather than inert end products. Bough and Gander (1971) collate ideas of Blumenthal *et al.* (1968) and Wright *et al.* (1958) suggesting that the cyanogenic glucoside dhurrin acts as a limited precursor in asparagine or lignin biosynthesis respectively. Asparagine is not the only possible product because in some species the ability to assimilate HCN is associated with the formation of β-cyanoalanine (*E. coli*; Dunnill and Fowden, 1965), γ-glutamyl-β-cyanoalanine (*Vicia sativa* L.; Fowden and Bell, 1965) or both (*Chlorella pyrenoidosa* Chick; Fowden and Bell, 1965) from cysteine (see Fig. 3). Elegant as this picture is, there are further difficulties. Does the availability of the CN⁻ of the cyanogenic glucoside depend upon the presence of the appropriate β-glucosidase? Is free HCN formed in the living tissue? Where is the β-glucosidase stored in the cell?

The first of these problems could be approached by studying the synthesis of asparagine from valine in plants which do not contain the appropriate β-glucosidase. Although there are plants which do not contain β-glucosidase in such concentration that it can hydrolyse the cyanogenic glucosides of *Lotus corniculatus* and *T. repens*, in all the tests which I have done, I still find residual activity for the *p*-nitrophenyl-β-D-glucopyranoside which is the synthetic substrate used for estimating the enzyme. Lloyd (1969) sensibly calls the enzyme involved *p*-nitrophenyl-β-D-glucopyranosidase. Blumenthal *et al.* (1968) have discussed the significance of the assimilation of HCN by several plants and certainly they argue that the conversion of HCN to asparagine could be a

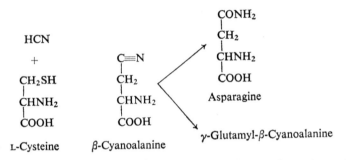

FIG. 3. Alternative pathways for the metabolism of HCN in some Leguminosae (Dunnill and Fowden, 1965)

detoxification process. Yet several species, which do not contain cyanogenic substances in detectable quantities, still possess the appropriate enzyme systems for converting HCN to asparagine. This suggests that HCN could well be present in these species, but not necessarily derived from cyanogenic glycosides. Feeding these acyanogenic plants with labelled amino acids known to be precursors of cyanogenic glycosides (Uribe and Conn, 1966) did not indicate, however, that HCN derived from cyanogenic glycosides was used as an intermediate stage in the synthesis of asparagine. Indeed the evidence suggested that only isoleucine of the four amino acids administered was concerned in any significant way. This paradox is unfortunate because, if HCN were commonly used for the synthesis of asparagine, then a ready explanation is available for the widespread distribution of cyanogenic glycosides in flowering plants. Other explanations (discussed later in this chapter) would be required for the discontinuity of distribution, but the comparability of the synthesis of, for example, linamarin in different species (Conn and Butler, 1969) could indicate homology between the genetic systems involved.

A lead on the third problem mentioned above is suggested by work on rat

110 DAVID A. JONES

liver. Patel and Tappel (1969) have shown that 80% of the β-glucosidase in the rat is attached to the lysosomal membrane. In addition Lloyd (1969) found that while the β-glucosidase is not on the cytoplasmic face of the lysosome, the membrane itself acts as a barrier to penetration by p-nitrophenyl-β-D-glucopyranoside. Similar studies with cyanogenic plant material would be more than useful, but there is circumstantial evidence on this.

Firstly there is evidence that hydrolytic enzymes, including α-glucosidases, occur in plant lysosomes (Matile, 1968). Secondly, it was pointed out earlier that the plant must be damaged before cyanogenesis begins. In an intact cell, therefore, the cyanogenic glucosides and the β-glucosidases must be kept apart chemically if not physically. Thirdly, Daday (1965) has argued that at low temperatures the β-glucosidase of *T. repens* is irreversibly activated. Some recent experiments subjecting *Lotus corniculatus* to a series of temperatures below 0°C, indicate that ice formation, beginning at −2°C, causes damage to the tissue of leaves and forms characteristic ice cracks (Taylor, personal communication). We infer that there is a relation between HCN production and the extent of ice formation because it is possible to supercool leaves without damaging them and without releasing HCN. This effect of partial freezing with release of HCN only on subsequent raising of the temperature is typical of the type of effect observed when lysosomes are damaged (De Duve, 1969; Matile, 1969).

V. PHYSIOLOGICAL CHARACTERISTICS OF PARASITES AND
GRAZERS OF CYANOGENIC PLANTS

Casual references to poisoning of large mammals by HCN (Henry, 1938) and formal investigations of poisoning of cattle (Prodanov and Zkelezova, 1962; Saad and Camargo, 1968) occur frequently in the literature. An enormous amount of work on the possible effects of cyanogenesis in white clover (*T. repens*) on livestock has been reported from New Zealand and the United States. Coop and Blakley (1949), for example, have shown that the minimum lethal dose of free HCN in sheep is approximately 2·4 mg HCN/Kg body weight. This is equivalent to 4·5 ± 0·5 mg of HCN/Kg produced by lotaustralin in one dose. On the other hand, they also have evidence that animals grazing normally and continuously are capable of tolerating a daily intake equivalent to 15–50 mg HCN/Kg. Sheep appear to respond to the effects of mild HCN poisoning by ceasing to feed until the symptoms have cleared. Rose (1941) has found that cattle could be maintained continuously and apparently with impunity on Sudan grass (*Sorghum vulgare sudanese* Hitchc.) containing as much as 0·113% HCN. This can be estimated as the equivalent of a daily intake of 50 mg/Kg. And yet starved cattle, offered Sudan grass for an hour or so each day as their only fresh food, suffered heavy mortality. A hungry cow eating 3·0 Kg of dry matter in 15 minutes would take in the equivalent of 6·0 mg HCN/Kg as Sudan grass.

It is clear that a large single dose can be fatal whereas the equivalent quantity of HCN spread over a longer period will produce only mild symptoms of cyanide poisoning or, indeed, no symptoms at all. Evidence is accumulating to show that other grazers and parasites of cyanogenic plants and animals are tolerant of HCN. Fry and Millar (1968) have observed that the fungus *Stemphylium loti* Graham causes cyanogenesis in *L. corniculatus* and that it is more tolerant of HCN than *S. sarciniiforme* (Cav.) Wiltshire which is a non-pathogen of bird's-foot trefoil. As mentioned previously, Jones *et al.* (1962) found that all stages of the life cycle of the burnet moths *Zygaena* spp., whose larvae have *L. corniculatus* as a food plant, contain cyanogenic substances. Larvae raised on acyanogenic plants which do not contain linamarin or lotaustralin still contain cyanogenic substances, so one must conclude that these are synthesized *de novo* by the larvae. Furthermore, we noted that a specific parasite, *Apanteles zygaenarum* Marshall, of these moths possesses a detoxifying enzyme which resembles rhodanese, whereas *A. tetricus* (Reinhard), a parasite of the non-cyanogenic butterfly *Maniola jurtina* L., does not.

The non-cyanogenic insect *Polyommatus icarus* Rott. shows no selective laying of eggs on *L. corniculatus* (Lane, 1962) and this is probably because the larvae contain the detoxifying enzyme rhodanese (Parsons and Rothschild, 1964). The weevil *Hypera plantaginis* Degeer which lives in the flower heads of the plant is also able to detoxify HCN. It seems likely that in these species the fortuitous acquisition (by mutation) of the ability to be tolerant of HCN has enabled the animal to move onto an unusual food plant and so reduce the competition with other species. For brief discussions of competition of this type in animals see Williamson (1957) and Ayala (1971).

Detoxification processes in mammals are also related to this enzyme rhodanese. It occurs predominantly in the liver (Lang, 1933) yet it is widely distributed in the body (Himwich and Saunders, 1948). The detoxification is aerobic, requiring a sulphur donor. Sodium thiosulphate is a clinical antidote to cyanide poisoning [usually administered intravenously with sodium nitrate (Chen *et al.*, 1944)]. The chemical reaction is:

$$CN^- + Na_2S_2O_3 = CNS^- + Na_2SO_3$$

and it seems likely that it is the normally low availability of sulphur which limits the efficiency of the detoxification process in the natural state. It has been shown by Meister (1953) and Wood and Fiedler (1953) that β-mercaptopyruvic acid is converted to pyruvic acid and free sulphur by extracts of liver and that this sulphur is available for the conversion of CN^- to CNS^- (See the review by Williams, 1959). Although the use of rhodanese is a common way of converting CN^- to CNS^- this process can be carried out by other means. For example, the xanthine oxidase in cow's milk is inactivated by cyanide, and this process is associated with the release of sulphur from the protein on the formation of CNS^- (Massey and Edmondson, 1970).

In extensive trials lasting a period of 14 months, Corkill (1952) obtained no evidence that sheep showed any preference for acyanogenic as opposed to cyanogenic *Trifolium repens*. It seems likely that the possession of a detoxification mechanism related to rhodanese has enabled sheep to be indiscriminatory with respect to cyanogenic white clover. There is a 200-fold change in toxicity on the conversion to CNS^- and it is likely that the process is a natural one because any cyanide produced during normal metabolism or produced from the diet of herbivores can be detoxified.

Several plant pathologists have investigated the possibility that cyanogenic substances afford protection against fungal attack. Up to now there is very little evidence in favour of that hypothesis. Trione (1960), for example, found no correlation between resistance to flax rust (*Fusarium oxysporum lini* (Bolley) Snyder et Hansen) and HCN content of the flax plant. It would not be surprising if this fungus contained a detoxifying mechanism. Treub (1896) argues that because cattle eat cyanogenic plants, and that the leaves of *Prunus japonica* Thunb. and the leaves of the upper branches of *Pangium edule* Reinw. containing HCN seem even to attract leaf eating parasites, then protection is not the role of HCN in these plants. Furthermore, Herissey (1923) considers that HCN production is too inefficient for cyanogenic glycosides to be considered as protective substances. The major difficulty with all these arguments against a protective role for cyanogenic glycosides is that one cannot rule out the possibility that some fungal or other parasites would attack the plants if they were not cyanogenic. It is necessary to examine polymorphic host plants to determine whether there is any differential parasitism (Jones, 1971). Millar and Higgins (1971) appear to have the right materials for studying the interaction between several species of fungi and polymorphic *Lotus corniculatus* but the critical experiment is still outstanding.

A most unusual association between parasite and plant host occurs with *Medicago sativa* L. (alfalfa) and the unidentified Basidiomycete parasite of the plant which causes winter crown rot. Some strains of the fungus contain an enzyme capable of causing cyanogenesis in the host (Colotelo and Ward, 1961) whereas other strains are capable of generating HCN autolytically without any requirement for the presence of host tissue or for the supply of specific substrates (Ward and Lebeau, 1962). The fungus can metabolize satisfactorily while releasing so much HCN into the medium; that which passes through the gaseous phase can be enough to kill alfalfa seedlings in separate liquid but in the same air space (Lebeau and Dickson, 1953). It can also be enough to kill dormant or vegetatively active plants whose root systems are in a sealed mass of soil which has been inoculated with the fungus. The pathological damage is greatest at low temperatures, and in the field its serious expression normally occurs when the snow melts in the spring—hence its vernacular name "Snow Mold" (Lebeau and Dickson, 1955). This feature of enhanced damage at low temperatures is consistent with certain other examples of cyanide damage, perhaps because of a greater solubility of HCN (Lebeau and Dickson, 1955) or due to a slower loss of HCN to the gaseous phase from water or cell sap at these low temperatures. Recent work shows that the cyanogenic substance in the snow mold (Ward and Thorn, 1966) and, indeed, that of the *Chromobacterium violaceum* (Schraeter) Bergonzini group as well (Michaels *et al.*, 1965) is derived from glycine.

Even though several animals and plants are able to eat or parasitize cyanogenic plants, this does not detract from the possible basic function of cyanogenesis as a defence mechanism. No defence mechanism is absolute and so we can consider cyanogenesis only in comparative and not in absolute terms as a defensive character. It may not be efficient, but it may well be sufficient to deter many would-be grazers and parasites.

VI. Genetical and Ecological Aspects of Cyanogenesis

Where the same character occurs in all members of one species, but in none of a closely related one, it is not possible to avoid confounding the character with the species. Indeed this state of affairs is ideal for the taxonomist. Yet there is little point studying species in which cyanogenesis is a monomorphic

Plant contains	Short hand notation	Gross phenotype	Varietal type
Cyanogenic glucosides and enzyme	G + E	Cyanogenic	Amara
Cyanogenic glucosides but no enzyme	G + noE	Acyanogenic	Dulcis
Enzyme but no cyanogenic glucosides	noG + E	Acyanogenic	Dulcis
Neither cyanogenic glucosides nor enzyme	noG + noE	Acyanogenic	Dulcis

FIG. 4. The phenotypes of cyanogenic and acyanogenic *Lotus corniculatus*

character when the object is to determine the role of cyanogenesis in natural populations. The ecological geneticist tackles this problem by studying the differential survival or reproductive value of different forms of the same species. This is a far more useful approach, because the theoretical models which apply suggest that a balance of selective forces will be involved in the maintenance of a genetical polymorphism (Ford, 1940, 1964, 1965). Suitable species for the study of the polymorphism of cyanogenesis are *Lotus corniculatus*, *Prunus amygdalus*, *Sorghum vulgare* Pers. and *Trifolium repens* in which both *dulcis* and *amara* forms occur. Because *L. corniculatus* and *T. repens* are commonly found in wild populations, these species have been used preferentially for this work.

Although cyanogenic plants contain both cyanogenic glycosides and the appropriate β-glucosidase, plants which are acyanogenic in gross phenotype may differ in their cyanogenic glycoside and β-glucosidase composition (Fig. 4). A similar scheme can be drawn up for *P. amygdalus*, *S. vulgare*, *T. repens* and probably for *Sambucus nigra* L. There is abundant evidence (see De Waal, 1942) that there is no correlation between morphology and HCN production in *T. repens* and so it is not possible to state that any particular plant is cyanogenic merely by visual inspection. Chemical tests are essential. This point needs to be clarified, however, for *L. corniculatus*.

In *T. repens* (Williams, 1939; Corkill, 1942; Atwood and Sullivan, 1943) and *L. corniculatus* (Dawson, 1941; Bansal, 1966) the presence of both the cyanogenic glucosides linamarin and lotaustralin is determined by a single dominant allele, while the presence of the appropriate β-glucosidase is also determined by a single dominant allele at a locus not genetically linked to the glucoside one. With *L. corniculatus* there is the added complication that the plant behaves as an autotetraploid for these loci (Dawson, 1941; Bansal, 1966).

De Waal (1942) spends much time arguing the case against the discontinuity between cyanogenic and acyanogenic plants being real. He is supported by Flück (1963) who suggests that the apparent absence of a substance may merely mean that the resolution of the analytical techniques is not fine enough to detect it in minute quantities. This attitude is verging on the metaphysical. As far as metabolism is concerned, minute traces of a compound may well be sufficient for life processes, but there is no doubt that a slug can distinguish between plants which picrate papers indicate are cyanogenic and acyanogenic, and these are the phenotypes which we are considering.

Because it is possible to find cyanogenic and acyanogenic plants of the same species, cyanogenesis is not necessary for the survival of the plant under all the normal conditions in which it can be found. Hence we are not dealing with characters which have absolute value to the species. This is not to say that the character is superfluous, as Conn and Butler (1969) appear to suggest, but merely that the relative advantage of the character to an individual will vary from place to place and habitat to habitat. The function of the cyanogenic glycoside in these species is, therefore, likely to be related to differences between certain components of habitats. If we can find out what these differences are, then we are well on the way to determining the selective agents involved in the maintenance of the polymorphism.

Two selective agents have been invoked; temperature by Daday (1954a, 1965) and selective eating (now to be called differential eating) by Jones (1962). Daday (1954a) obtained evidence that there is a relationship between the frequency of the cyanogenic form of *T. repens* in natural populations and the January mean temperature (Fig. 5). More convincing is the relation between cyanogenesis and altitude, as can be seen from Fig. 6 (Daday, 1954b), but it was not until 1965 that Daday showed by direct experimentation that the acyanogenic NoG + NoE form is at a selective disadvantage in New South Wales where the July mean temperature is 13°C, whereas it is at an advantage in an alpine environment with a July mean temperature of -2°C. He suggests that it is not the cyanogenic character *per se* which is responsible for the advantage of cyanogenic plants at high winter temperature, but that the locus concerned with cyanogenic glucoside production is genetically linked to genes concerned with fitness responses to temperature. At low temperature, frost would have a direct effect by causing cyanogenesis in cyanogenic plants.

Daday has been forced to adopt this explanation because he found that plants which contain the cyanogenic glucosides but no β-glucosidase (i.e.

G + NoE) were as fit, biologically, as cyanogenic plants, yet G + NoE plants are acyanogenic in phenotype and so temperature cannot be having a direct effect on fitness via the character of cyanogenesis. When cyanogenesis shows obvious disadvantages under frost conditions and acyanogenic plants can be

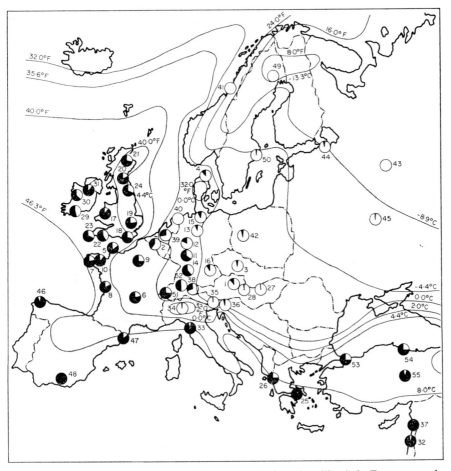

FIG. 5. Distribution and frequency of the cyanogenic form (see Fig. 4) in European and Near Eastern wild populations of *Trifolium repens* L. Black section: frequency of the cyanogenic form. White section: frequency of the acyanogenic forms. —— January isotherm. [Modified after Daday (1954a) by permission of the editor and publishers of *Heredity*]

equally as fit as cyanogenic ones, one wonders why cyanogenesis remains in the population? It is more likely that the phenotype of cyanogenesis itself is selected directly and this is the rationale behind the approach which Jones (1970) has taken.

There were at least two records of differential eating of acyanogenic plants before Jones (1962) reported a formal investigation of selection of the acyanogenic form of *Lotus corniculatus* by several species of slug and snail. Corkill (1952) and Daday (1955) observed it in their experimental plots and both Barber (1955) and Morley (1959) comment on these observations. All along, I have been impressed by the correlation between the frequency of occurrence of the cyanogenic glucosides and the β-glucosidase in population samples (Table II) and I have argued that cyanogenesis is the main purpose of this enzyme-

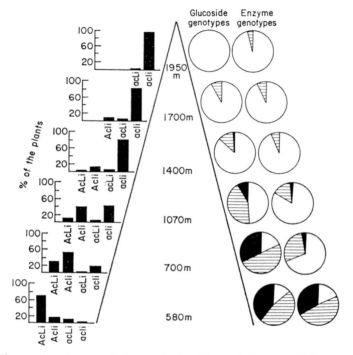

FIG. 6. Phenotypic and genotypic frequencies in wild populations of *Trifolium repens* from different altitudes

Phenotypes (left):   Estimated Genotypes (right):

AcLi = glucosides and enzyme   Black section = dominant homozygotes
Acli = glucosides only   Lined section = heterozygotes
acLi = enzyme only   White section = recessive homozygotes
acli = neither glucosides nor enzyme

(Modified from Daday, 1954b, with the permission of the editor and publishers of *Heredity*.)

substrate system. The initial observations (Jones, 1962) on selective eating were made in natural populations and, subsequently, formal experiments were devised (Jones, 1966). These latter showed conclusively that the acyanogenic form of *Lotus corniculatus* was preferentially eaten when a choice was offered to several species of slug and snail. Bishop and Korn (1969), however, did not obtain evidence of selective eating of *T. repens* by the slug *Agriolimax reticulatus* Müll. and the snail *Helix aspersa* Müll. Crawford-Sidebotham (1971), on the

TABLE II

Analysis of variance of the frequency of the enzymatic form on the frequency of the glucoside form of *Trifolium repens* [data of Daday, 1954a] and of *Lotus corniculatus*, [data of Jones, 1970. Reproduced by permission of the editor and publishers of *Heredity*].

		d.f.	M.S.	
T. repens	Regression	1	18123·42	$P < 0.001$
	Residual	47	83·45	
L. corniculatus	Regression	1	6914·73	$P < 0.001$
	Residual	12	158·12	

other hand, has obtained results summarized in Table III, which show that, in general, where a species of mollusc shows differential eating for one plant species only, it scarcely eats the other species at all. There is, however, one anomaly which has yet to be explained. *Agriolimax reticulatus* eats *T. repens* whether the plant is cyanogenic or acyanogenic, yet it shows differential eating of *L. corniculatus*.

TABLE III

Summary of differential eating experiments with 13 species of slug and snail (from Crawford-Sidebotham, 1971).

	Lotus corniculatus	*Trifolium repens*
No. of species† showing preference for the acyanogenic form	7*	7*
No. of species† showing no selection or no eating of the plant	6	6

* These numbers do not contain exactly the same species; see text.
† Slugs and snails feeding on both plants: *Agriolimax reticulatus* Müll., *Arion ater* (L.), *A. hortensis* Fér., *Cepaea hortensis* (Müll.), *C. nemoralis* (L.), *Helix aspersa* Müll. and *Monacha cartusiana* (Müll.); feeding only on *T. repens*: *Arianta arbustorum* (L.) and *Helicella virgata* (da Costa); feeding only on *L. corniculatus*: *Arion subfuscus* (Drap.); no grazing on either plant: *Agriolimax caruanae* Poll; *Milax budapestensis* (Hazay) and *Theba pisana* (Müll.).

Within each molluscan species, however, the data are heterogeneous and we conclude that, in the same way that some men like beer and others do not, individual molluscs have different palates. Crawford-Sidebotham confirmed the lack of selective eating of *T. repens* by *Agriolimax reticulatus*, but he obtained two results at variance with those previously published. No significant eating of either form of *L. corniculatus* by the snail *Arianta arbustorum* (L.)

was observed, whereas *Helix aspersa* did prefer acyanogenic *T. repens*. There is another way of accounting for differences between the behaviour of individuals. It is possible that a mollusc will ignore a cyanogenic plant, not because it does not like the taste, but because it is not hungry. All the animals used by Crawford-Sidebotham were subjected to a period of forced fasting before an experiment and so it is likely that the animals were hungry. Indeed, some animals preferred to eat the filter paper in the dish than the leaves.

What has not been done in anything other than a tentative way, is to relate the distribution of animals known to eat acyanogenic plants differentially with the distribution of the plant morphs in natural populations. It is likely that the selective effect observed on mature plants is too coarse to be of particular importance in established populations, but it is an entirely different matter with seedlings. When grazing by animals is considered it is clear that the most sensitive stage of plant growth is the young seedling. Thus any mechanism which protects a seedling will be at a high selective advantage. Very frequently reference is made to acyanogenic seeds showing rapid synthesis of cyanogenic glycosides on germination. This process occurs in *Manihot utilissimus* (Nartey, 1968) where linamarin (93%) and lotaustralin (7%) are synthesized, in *Sorghum vulgare* (Bough and Gander, 1971) in *Lotus corniculatus* (Jones and Crawford-Sidebotham, unpublished) and in related species (Hegnauer, 1961). It would appear, therefore, that cyanogenesis in seedlings could be a means whereby the young plants are protected from grazing and phytophagous animals. It is likely that any effects would be density dependent.

If the presence of cyanogenic glucosides in *T. repens* and *L. corniculatus* is determined by some special metabolic requirement, or indeed if they are gratuitous or waste products there should be no interactions between the two species with respect to cyanogenesis. That is, the frequency of cyanogenic forms in, say, *T. repens* in a given habitat type should be the same whether *L. corniculatus* is there or not. Jones (1968) has, however, recorded that in the 8 mixed colonies he studied, the frequency of the cyanogenic form of *T. repens* was invariably lower than the frequency of that form in *L. corniculatus*. Moreover, amongst groups of *T. repens* plants growing separately from *L. corniculatus*, but within 25 m of the mixed groups, there was a significantly higher frequency of the cyanogenic form in four of the populations.

Thus there is evidence of an interaction between the two species and thus a further indication that the *phenotype* of cyanogenesis is significant in these species.

VII. Discussion and Conclusions

A. Cyanogenic Glycosides as Waste Products in Plants

The isolation of the cyanogenic glycoside prunasin from the bark of *Prunus serotina* Ehrh. (Power and Moore, 1909) and the frequent occurrence of cyanogenic substances in external regions of old stems suggested to Goris (1921) that, because plants may have difficulty in bringing about the breakdown of cyclic compounds, these could be stored in an innocuous form as cyanhydrin glycoside. It did not occur to him that deposition of these substances in bark could equally well result in the reduction of bark-grazing in winter by deer and rabbits. Goris further suggested that these substances could be regarded as by-products of cell activity, because there are several examples of cyanogenic glycosides being absent from seeds but appearing in the early stages of germination.

Robinson (1930) has argued against this interpretation by pointing out that the greatest economy is exhibited by plants with respect to nitrogen. In addition there is some evidence (Mayer and Poljakoff Mayber, 1963) that nitrogen is very carefully conserved in germinating seeds.

Treub (1896) has shown that in *Panguim edule* the concentration of cyanogenic substances diminishes with the age of the leaves and they disappear altogether by leaf fall. Combes (1918) declares that substances which are regarded as excretions are really those whose real function is unknown. Combes also warns against studying the effects of natural products on plants which do not synthesize them and then drawing conclusions about the part they play in the tissues of species which in nature do produce them.

None of the authors has considered the presence of cyanogenic glycoside in seeds as being in any way significant, yet bitter almonds can contain as much as 0·25% amygdalin. It is most unlikely that a seed will contain waste products at such a concentration.

B. CYANOGENIC GLYCOSIDES AND AMINO ACID METABOLISM

It was Treub (1896) who argued that the presence of cyanogenic substances in the phloem of *Pangium edule* indicated its importance as a translocatory material. Various references have been made in the past to the disappearance of, or a reduction in, the cyanogenic glucoside content of plants on cultivation. One view of this (Dunstan and Henry, 1903) is that the improved conditions of nutrition and environment in cultivation cause a stimulus to metabolism leading to a more rapid use of plastic substances, so that accumulation is less obvious or non-existent. Although there is no long term evidence on this point, De Waal (1942) noted, on the contrary, that the linamarin content of *Trifolium repens* rose when nitrogen manure was supplied to individual plants in plots.

Certainly Abrol *et al.* (1966) have shown active turnover of cyanide in *Lotus* spp. and *Nandina domestica*. Similarly Tschiersch (1966) has shown that the cyanogenic substance vicianin is likely to be an intermediate between phenylalanine -2-^{14}C and β-cyanoalanine in *Vicia angustifolia* L. It is the biosynthesis of asparagine, however, which is of most interest because there is no well characterized enzyme for asparagine biosynthesis from aspartic acid, as there is for the step glutamic acid to glutamine (Fowden, 1967), and so the synthetic pathway using HCN is attractive. As was explained earlier, Blumenthal *et al.*

(1968) have done some experiments with *Lupinus angustifolius* (completely acyanogenic) and *Vicia angustifolia* (synthesizes β-cyanoalanine from HCN). They have failed to show that the asparagine formed in experiments with these species using ^{14}C labelled phenylalanine, tyrosine, valine and isoleucine carries the asymmetric distribution of the ^{14}C label expected if HCN were an intermediate. Conn and Butler (1969) conclude that "the role for cyanide assimilation is one of detoxification which is useful in the case of cyanophoric plants. This may be a metabolic activity acquired early in evolution and retained by species which no longer have a need for such a process". It may be that these non-cyanogenic plants which can metabolize HCN have HCN available from sources other than cyanogenic glycosides and this could be the reason why the ability is retained. Although it would appear that some species metabolize HCN and by that means detoxify this compound, there is evidence that other cyanogenic plants, *Arum maculatum* L. for example, have evolved respiratory systems which are not completely inhibited by HCN (see Schonbaum *et al.*, 1971 for references).

C. CYANOGENIC GLYCOSIDES AND DIFFERENTIAL EATING

The major embarrassment with the hypothesis of differential eating of acyanogenic forms is that it is so eminently reasonable. Whether these substances occurred originally as waste products or intermediates in metabolism cannot be determined, but there can be little doubt that the concentration of cyanogenic substances in many plants is considerably higher than would be needed for normal metabolism. That plants of the same species can be either cyanogenic or acyanogenic is good evidence that this is so.

Differential eating has now been adequately demonstrated using many species of mollusc and two polymorphic species of plant, and therefore we are dealing with a real phenomenon. The interaction between the two species—i.e. the effect of cyanogenic *Lotus corniculatus* on the frequency of that form in *T. repens*—clearly demonstrates that a purely metabolic role for cyanogenic glycosides in these plants must be ruled out. The enzymes which convert HCN to asparagine or β-cyanoalanine may well be part of a detoxification process used after HCN has been released by a deterred grazer or pathogen.

The temperature effects alone, on the other hand, do not provide a satisfactory alternative explanation of the selective effects because they do not explain the advantage which cyanogenic plants have at high temperature. It is more likely that both selective agents are important, probably in an oscillating way (Jones, 1970).

A scheme which outlines the likely evolutionary sequence for cyanogenesis is contained in Fig. 7. The principle is not entirely novel because similar models have been presented elsewhere (Flor, 1956; Person *et al.*, 1962). The details are new and have been referred to in other parts of this review. The diagram explains the situation where the cyanogenic form becomes fixed. Species which are polymorphic at the present time may (a) be at the transient stage, or (b) have achieved a balanced polymorphism [probably clinal as in *T. repens* (Daday, 1954a)], where a balance has been achieved between metabolic cost and natural selection.

There is a major problem still outstanding. *L. corniculatus* and *T. repens* are closely related species and the synthetic pathways leading to the production of the cyanogenic glucosides, linamarin and lotaustralin, are identical as far as one can tell in the two species (Abrol and Conn, 1966). Yet there are acyanogenic species within each genus. Why are not all *Lotus* and *Trifolium* species

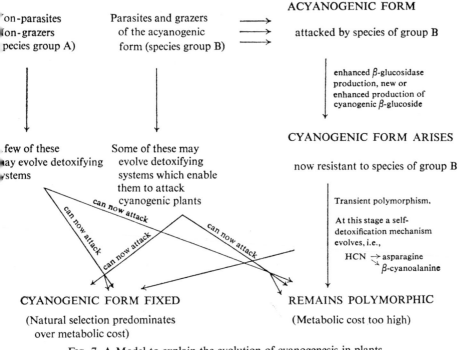

cyanogenic, when, as the differential eating experiments suggest, cyanogenesis acts as a protective mechanism? Perhaps these other species use alternative defence mechanisms which have not yet been identified.

In all this discussion, we are continually brought back to the phenotype of cyanogenesis and clearly this is the starting point of all argument. It is here, I suspect, that differences of opinion between biochemist and ecological geneticist will be most marked, but certainly it is the experience of the geneticist that if qualitative and/or quantitative differences between individuals can be shown to have different selective values, then there is no doubt that the phenotype, as observed, is important for the survival of individuals, of the genotype and, perhaps, even of the species.

ACKNOWLEDGEMENTS

I am grateful to the Science Research Council for financial support of some of the work reported here and to Dr T. J. Crawford-Sidebotham for his enthusiasm and constructive comments on this work. I also thank Professor J. A. Beardmore who translated De Waal's thesis from the original Dutch. It is a pleasure to pay my respects to Professor E. B. Ford on the occasion of his 70th birthday.

REFERENCES

Abrol, Y. P. and Conn, E. E. (1966). *Phytochemistry* **5**, 237.
Abrol, Y. P., Conn, E. E. and Stoker, J. R. (1966). *Phytochemistry* **5**, 102.
Askew, H. O. (1933). *N.Z. Jl Sci. Tech.* **15**, 227.
Atwood, S. S. and Sullivan, J. T. (1943). *J. Hered.* **34**, 311.
Ayala, F. J. (1971). *Science, N.Y.* **171**, 820.
Bansal, R. D. (1966). "Studies on Procedures for combining clones of birdsfoot trefoil, *Lotus corniculatus* L." Ph.D. Thesis, Cornell University, U.S.A.
Barber, H. N. (1955). *Austr. J. Sci.* **18**, 148.
Barman, T. E. (1969). "Enzyme Handbook", p. 578, Springer Verlag, Heidelberg.
Berti, G. and Bottani, F. (1968). *In* "Progress in Phytochemistry" (L. Reinhold and Y. Liwschitz, eds), Vol. I, pp. 589–685. Interscience, New York.
Bishop, J. A. and Korn, M. E. (1969). *Heredity* **24**, 423.
Blumenthal-Goldschmidt, S., Butler, G. W. and Conn, E. E. (1963). *Nature, Lond.* **197**, 718.
Blumenthal, S. G., Hendrickson, H. R., Abrol, Y. P. and Conn, E. E. (1968). *J. biol. Chem.* **243**, 5302.
Bough, W. A. and Gander, J. E. (1971). *Phytochemistry* **10**, 67.
Butler, G. W. (1965). *Phytochemistry* **4**, 127.
Butler, G. W., Bailey, R. W. and Kennedy, L. D. (1965). *Phytochemistry* **4**, 369.
Butler, G. W. and Butler, B. G. (1960). *Nature, Lond.* **187**, 780.
Chen, K. K., Rose, C. L. and Clowes, G. H. A. (1944). *J. Indiana med. Ass.* **37**, 344.
Clawson, B. J. and Young, C. C. (1913). *J. biol. Chem.* **15**, 419.
Colotelo, N. and Ward, E. W. B. (1961). *Nature, Lond.* **189**, 242.
Combes, R. (1918). *Rev. gen. Bot.* **30**, 355.
Conn, E. E. and Butler, G. W. (1969). *In* "Perspectives in Phytochemistry" (J. B. Harborne and T. Swain, eds), pp. 47–74. Academic Press, London and New York.
Coop, I. E. and Blakley, R. L. (1949). *N.Z. Jl Sci. Tech.* **30A**, 277.
Corkill, L. (1942). *N.Z. Jl Sci. Tech.* **23B**, 178.
Corkill, L. (1952). *N.Z. Jl Sci. Tech.* **34A**, 1.
Crawford-Sidebotham, T. J. (1971). "Studies of aspects of slug behaviour and the relation between molluscs and cyanogenic plants". Ph.D. Thesis, University of Birmingham.
Daday, H. (1954a). *Heredity* **8**, 61.
Daday, H. (1954b). *Heredity* **8**, 377.
Daday, H. (1955). *J. Br. Grassland Soc.* **10**, 266.
Daday, H. (1962). *Ann. Rep. Div. Plant Ind., C.S.I.R.O. Canberra*, 1961–1962, XIII and 14.
Daday, H. (1965). *Heredity* **20**, 355.
Dawson, C. D. R. (1941). *J. Genet.* **42**, 49.
Duerksen, J. D. and Halvorson, H. O. (1959). *Biochim. biophys. Acta* **36**, 47.
Dunnill, P. M. and Fowden, L. (1965). *Nature, Lond.* **208**, 1206.

Dunstan, W. R. and Henry, T. A. (1903). *Proc. Roy. Soc.* **72**, 285.
De Duve, C. (1969). *In* "Lysosomes in Biology and Pathology" (J. T. Dingle and H. B. Fell, eds). North Holland, Amsterdam.
De Waal, D. (1942). "Het Cyanophore Karakter van Witte Klaver," Ph.D. Thesis, Agricultural University, Wageningen, The Netherlands.
Ehrlich, P. R. and Raven, P. H. (1965). *Evolution* **18**, 586.
Eisner, H. E., Eisner, T. and Hurst, J. J. (1963a). *Chemy Ind.* 1963, 124.
Eisner, T., Eisner, H. E., Hurst, J. J., Kafatos, F. C. and Meinwald, J. (1963b). *Science, N.Y.* **139**, 1218.
Flor, H. H. (1956). *Adv. Genet.* **8**, 29.
Flück, H. (1963). *In* "Chemical Plant Taxonomy" (T. Swain, ed.). Academic Press, London and New York.
Ford, E. B. (1940). *In* "The New Systematics" (J. S. Huxley, ed.), pp. 493–513. Clarendon Press, Oxford.
Ford, E. B. (1964). "Ecological Genetics." Methuen, London.
Ford, E. B. (1965). "Genetic Polymorphism." Faber and Faber, London.
Fowden, L. (1967). *Ann. Rev. Pl. Physiol.* **18**, 85.
Fowden, L. and Bell, E. A. (1965). *Nature, Lond.* **206**, 110.
Fraenkel, G. (1959). *Science, N.Y.* **129**, 1466.
Fraenkel, G. (1969). *Ent. Exp. App.* **12**, 473.
Fry, W. E. and Millar, R. L. (1968). *Phytopathology* **58**, 399.
Gibbs, R. D. (1963). *In* "Chemical Plant Taxonomy" (T. Swain, ed.), pp. 60–82. Academic Press, London and New York.
Goris, A. (1921). *Rev. Gen. Sci.* **32**, 337.
Guyot, H. (1916). *Bull. Soc. Bot. Genève* **8**, sér. 2, 80.
Haisman, D. R. and Knight, D. J. (1967). *Biochem. J.* **103**, 528.
Hahlbrock, K. and Conn, E. E. (1970). *J. biol. Chem.* **245**, 917.
Hahlbrock, K. and Conn, E. E. (1971). *Phytochemistry*, **10**, 1019.
Hahlbrock, K., Tapper, B. A., Butler, G. W. and Conn, E. E. (1968). *Archs Biochem. Biophys.* **125**, 1013.
Hegnauer, R. (1961). *Pharm. Weekbl.* **84**, 248.
Hegnauer, R. (1963). "Chematoxonomie der Pflanzen", Vols. I–IV. Berkhauser, Basel.
Henry, T. A. (1938). *J. Soc. Chem. Ind.* **57**, 248.
Hérissey, H. (1923). *Bull. Soc. Chim. Fr.* **33**, 349.
Himwich, W. A. and Saunders, J. P. (1948). *Am. J. Physiol.* **153**, 348.
Hutton, E. M. (1957). *J. Austr. Inst. Agric. Sci.* **23**, 227.
Jones, D. A. (1962). *Nature, Lond.* **193**, 1109.
Jones, D. A. (1966). *Can. J. Genet. Cytol.* **8**, 556.
Jones, D. A. (1968). *Heredity* **23**, 453.
Jones, D. A. (1970). *Heredity* **25**, 633.
Jones, D. A. (1971). *Science, N.Y.* **173**, 945.
Jones, D. A., Parsons, J. and Rothschild, M. (1962). *Nature, Lond.* **193**, 52.
Kingsbury, J. M. (1964). "Poisonous Plants of the United States and Canada." Prentice-Hall, Englewood Cliffs, New Jersey.
Lane, C. (1962). *Ent. Gaz.* **13**, 112.
Lang, K. (1933). *Biochem. Z.* **259**, 243.
Lebeau, J. B. and Dickson, J. G. (1953). *Phytopathology* **43**, 581.
Lebeau, J. B. and Dickson, J. G. (1955). *Phytopathology* **45**, 667.
Levin, D. A. (1971). *Am. Nat.* **105**, 157.
Lloyd, J. B. (1969). *Biochem. J.* **115**, 52P.
Lösecke, A. Von (1971). *Bolt. Arch. Pharm.* **147**, 36.
Mao, E. H. and Anderson, L. (1967). *Phytochemistry* **6**, 473.

Massey, V. and Edmondson, D. (1970). *J. biol. Chem.* **245**, 6595.
Matile, P. H. (1968). *Planta* **79**, 181.
Matile, P. H. (1969). In "Lysosomes in Biology and Pathology" (J. T. Dingle and H. B. Fell, eds). North Holland, Amsterdam.
Mayer, A. M. and Poljakoff-Mayber, A. (1963). "The Germination of Seeds." Pergamon Press, Oxford.
Meister, A. (1953). *Fedn. Proc.* **12**, 245.
Michaels, R., Hankes, L. V. and Corpe, W. A. (1965). *Archs Biochem. Biophys.* **111**, 121.
Millar, R. L. and Higgins, V. J. (1970). *Phytopathology* **60**, 104.
Morley, F. H. W. (1959). *Cold Sp. Harb. Symp. Quant. Biol.* **24**, 47.
Nartey, F. (1968). *Phytochemistry* **7**, 1307.
Paris, R. (1963). In "Chemical Plant Taxonomy" (T. Swain, ed.), pp. 337–358. Academic Press, London and New York.
Parsons, J. and Rothschild, M. (1964). *Ent. Gaz.* **15**, 58.
Patel, V. and Tappel, A. L. (1969). *Biochem. biophys. Acta* **191**, 86.
Patty, F. A. (1921). *J. Infect. Dis.* **29**, 73.
Person, C., Samborski, D. J. and Rohringer, R. (1962). *Nature, Lond.* **194**, 561.
Power, F. B. and Moore, C. W. (1909). *Chem. Soc. J.* **95**, 243.
Prodanov, P., and Zkelezova, B. (1962). *Bulgar. Akad. Nauk.* **9**, 285.
Robinson, D., Price, R. G. and Dance, N. (1967). *Biochem. J.* **102**, 533.
Robinson, Muriel E. (1930). *Biol. Rev.* **5**, 126.
Rogers, C. F. and Frykolm, O. C. (1937). *J. Agric. Res.* **55**, 533.
Rose, A. L. (1941). *Aust. Vet. J.* **17**, 211.
Saad, A. D. and Camargo, W. V. A. (1968). *Biologico (São Paulo)* **33**, 211.
Schrader, J. C. C. (1803). *Gilbert, Annalen* **13**, 503.
Schaefler, S. (1967). *J. Bact.* **93**, 254.
Schonbaum, G. S., Bonner, W. D., Storey, B. T. and Bahr, J. T. (1971). *Plant Physiol.* **47**, 124.
Seigler, D., Seaman, F. and Mabry, T. J. (1971). *Phytochemistry* **10**, 485.
Sharples, D. and Stoker, J. D. (1969). *Phytochemistry* **8**, 597.
Stahl, E. (1888). *Jena Z. med. Naturw.* **22**, 557.
Stevens, D. L. and Strobel, G. A. (1968). *J. Bact.* **95**, 1094.
Tapper, B. A. and Butler, G. W. (1971). *Biochem. J.* **124**, 935.
Treub, M. (1896). *Ann. Jard. Bot. Buitenzorg* **13**, 1.
Trione, E. J. (1960). *Phytopathology* **40**, 482.
Tschiersch, B. (1966). *Flora. Abt. A.* **157**, 43.
Uribe, E. G. and Conn, E. E. (1966). *J. biol. Chem.* **241**, 92.
Viebel, S., Jensen, K. and Klajan, E. (1963). *Biochem. Z.* **337**, 146.
Ward, E. W. B. and Lebeau, J. B. (1962). *Can. J. Bot.* **40**, 85.
Ward, E. W. B. and Thorn, C. D. (1966). *Can. J. Bot.* **44**, 95.
Whittaker, R. H. and Feeny, P. (1971). *Science, N.Y.* **171**, 737.
Williams, R. D. (1939). *J. Genet.* **38**, 357.
Williams, R. T. (1959). "Detoxification Mechanisms", pp. 390–409. Chapman and Hall.
Williamson, M. (1957). *Nature, Lond.* **180**, 422.
Wokes, F. and Willimott, S. G. (1951). *J. Pharm. Pharmac.* **3**, 905.
Wood, J. L. and Fiedler, H. (1953). *J. biol. Chem.* **205**, 231.
Wright, D., Brown, S. A. and Neish, A. C. (1958). *Can. J. Biochem. Physiol.* **36**, 1037.
Wöhler, F. and Liebig, J. (1837). *Ann. Chem.* **22**, 1.

CHAPTER 8

Aflatoxin and Related Mycotoxins

M. O. MOSS

*Department of Biological Sciences, University of Surrey,
Surrey, England*

I. Introduction

The large group of saprophytic organisms, the fungi, are associated with the production of a wide range of compounds which have excited the interest of the chemist, because of their chemical novelty, and of the biochemist, because of the problems that they pose as to their role and biosynthesis in microbial metabolism. Although there has long been an intrinsic interest in fungal metabolites, and one would like to pay tribute to such people as Professor Raistrick for much pioneer work in this field, further interest has been stimulated by the association of biological activity with many such metabolites. The number of useful antibiotics produced by the fungi is relatively small but those, such as the pencillins, cephalosporins and griseofulvin, which are produced industrially, have proved to be of such great value in the treatment of disease that this particular type of biological activity will continue to be a major stimulus to the study of mould metabolites.

Many fungal metabolites have been found to have antibiotic activity but, after further work with them, they have been rejected because of their toxicity to higher animals. Interest in fungal metabolites, because of this toxicity to man or domestic animals, is a relatively new phenomenon on an international scale. Previous to the beginning of the last decade there had been detailed studies of isolated examples of fungal toxins in particular parts of the world and these are described in some of the reviews on mycotoxins which are listed separately at the beginning of the bibliography. It is worth citing a few examples of these studies as an indication of the range of fungi and chemical structures involved.

The association of ergotism, or St. Anthony's Fire, with the metabolites of the ascomycete *Claviceps purpurea* led to a considerable study of these complex alkaloids (Fig. 1). In Russia the association of alimentary toxic aleukia with the toxic metabolites of a number of species of *Fusarium*, and of a disease of horses with *Stachybotrys atra*, led to the early establishment of studies of mycotoxins (Fig. 2). In Japan outbreaks of illness and death following the consumption of moulded rice led to detailed studies of what became known as

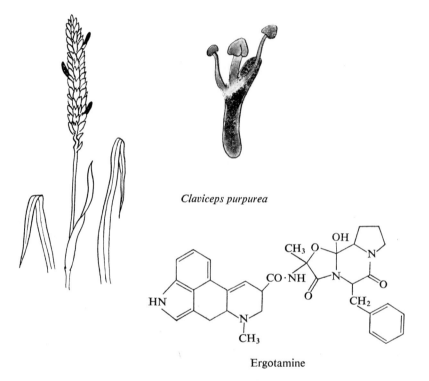

Claviceps purpurea

Ergotamine

FIG. 1. The head of a cereal infected with *Claviceps purpurea*, the dark sclerotia replacing some of the ears of the cereal. After falling to the ground and overwintering the sclerotia germinate to produce the ascomycete stage. Ergotamine is one of many complex alkaloids which can be isolated from the sclerotia

yellow rice disease. A number of species of the genus *Penicillium* and the toxic metabolites elaborated by them have been associated with this particular disease, one of the most important being *P. islandicum* and its metabolites islanditoxin and luteoskyrin (Fig. 3). In New Zealand the sheep farmers have had to contend with the continual possibility of a disease of their animals which they called facial eczema. An intensive study of the aetiology of this disease finally led to a single saprophytic member of the fungi imperfecti, *Pithomyces chartarum*, and its toxic metabolites the sporidesmins (Fig. 4).

II. AFLATOXINS

International interest in mould metabolites toxic to man and his domestic animals was awakened by an outbreak of disease amongst turkey poults in parts of England in 1960 (Blount, 1961). The name Turkey X disease suggests that the first thoughts on the nature of these deaths may have been that they were caused by a contagious factor such as a virus. It was soon shown that such was not the case and, because a highly intensive, economically important

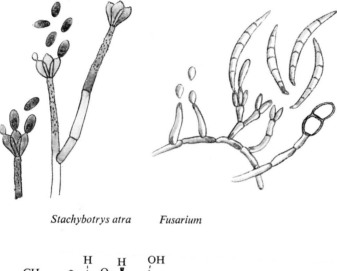

Stachybotrys atra   Fusarium

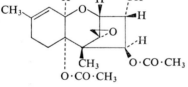

Diacetoxyscirpenol

Fɪɢ. 2. The dark unicellular phialospores of *Stachybotrys atra* are produced on well defined pigmented rough walled conidiophores whereas the colourless micro- and macrospores of many *Fusarium* species are produced from phialides borne on ill defined conidiophores. Diacetoxyscirpenol is a particularly virulent toxin associated with some strains of *Fusarium* but not necessarily implicated in alimentary toxic aleukia

industry was threatened, the aetiology of these outbreaks was studied with high priority. As a result of the work of the interdisciplinary teams involved in these early studies, the disease was soon associated with a particular mould— *Aspergillus flavus*—(Sargeant *et al.*, 1961) and its interesting metabolites the aflatoxins. There is now such a corpus of information on this organism and its metabolites that any individual can only cream off retrospectively those areas

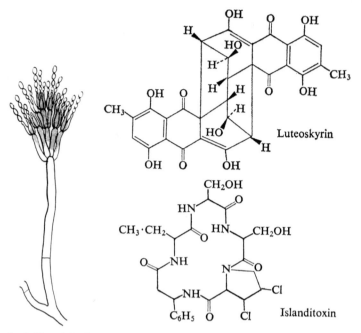

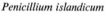

Penicillium islandicum

FIG. 3. One of the penicillia associated with yellow rice disease and two of its toxic metabolites

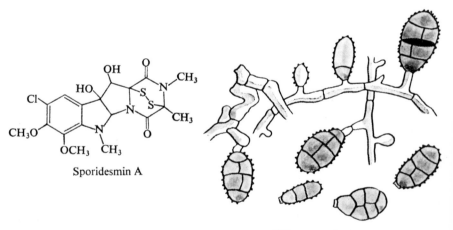

Pithomyces chartarum

FIG. 4. The brown aleuriospores of *Pithomyces chartarum* are formed as the blown out ends of inconspicuous hyaline conidiophores. Typically the spores take a small fragment of the conidiophore with them when they are released. Sporidesmin A is one of a number of the metabolites from this mould believed to be responsible for facial eczema of sheep

that particularly interest him and present them as an incomplete skeleton of the whole.

International interest in this one mould and its metabolites is reflected by the considerable number of publications appearing in the world literature since 1960 (Fig. 5).

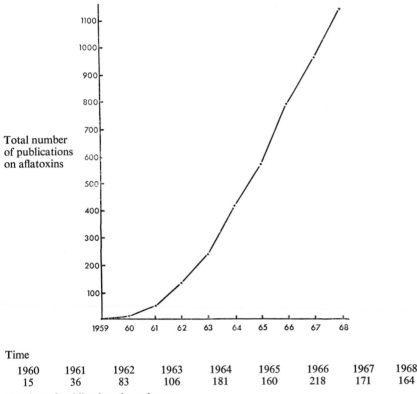

Time								
1960	1961	1962	1963	1964	1965	1966	1967	1968
15	36	83	106	181	160	218	171	164

Number of publications in each year

FIG. 5. The rate of increase of publications relating to aflatoxin. (Based on a Tropical Products Institute Report, 1967)

Before discussing the subject retrospectively it is worth indicating the time in which the subject has developed as it was reported in the literature:

1960—Turkey X disease was described.

1961—It was realized that other animal disease may be caused by the same agent as that causing Turkey X disease and this agent was shown to be a fungal metabolite, 20 μg of which is lethal to ducklings in 24 hours (Sargeant et al., 1961). It was also demonstrated that the same agent is carcinogenic, at least to rats (Lancaster et al., 1961).

6

1962—Some of the physico-chemical properties and analytical techniques for studying the aflatoxins were described. It was now realized that a family of compounds are involved and that they are photosensitive in hydroxylic solvents (Van der Zijden *et al.*, 1962).

1963—The chemical structures of the aflatoxins were described (Asao *et al.*, 1963).

1964—X-ray crystallographic confirmation of the structures was reported (Cheung and Sim, 1964).

1965—The aflatoxins were associated with the disturbance of events in the cell nucleus (Legator *et al.*, 1965).

1966—The total synthesis of racemic aflatoxin B_1 was described (Büchi *et al.*, 1966).

Aspergillus flavus (Link ex Fries) is not only a widespread saprophytic mould and an occasional pathogen of animals but it is very closely related to strains of *Aspergillus* used extensively in the production of foods and beverages in the Far East and other parts of the world (see Hesseltine, 1965). Thus strains of *A. flavus* itself and of *A. oryzae* are used in the manufacture of soy sauce and miso. A third species of the group, *A. parasiticus*, can be distinguished using morphological characteristics from those organisms used most frequently in the food industry, and it is with this species that the most consistently high yields of the aflatoxins are associated (Hesseltine *et al.*, 1970).

The aflatoxins (Fig. 6) are a family of compounds the first four of which are labelled B_1, B_2, G_1 and G_2 because of their characteristic blue and green fluorescence when viewed as spots on chromatograms in ultraviolet light. The presence of the fused difuran ring system attached to an already substituted coumarin structure makes these compounds very unusual and it is tempting to presume that they will only be produced by a limited range of organisms. There are reports in the literature of the production of aflatoxins by fungi other than *A. flavus* and its very close relatives but none has been widely substantiated and some have been specifically refuted. It has been claimed by Scott *et al.* (1967) that a strain of *A. ostianus* produces aflatoxins. Van Walbeek and his colleagues (1968) suggest that *A. ochraceous* and a species of *Rhizopus* produce them, and Hodges *et al.* (1964) report that strains of *Penicillium puberulum* also produce these toxins. It is known that unusual metabolites may be produced by different species even from different genera. The unusual di-isonitrile xanthocillin X (Fig. 7) is produced by both *A. chevalieri* (Coveney *et al.*, 1966) and *P. notatum* (Rothe, 1954) while the penicillin nucleus, a very singular chemical entity, is produced by representatives of a number of genera of fungi (see Cole, 1966).

Because of their potent biological activity at very low concentrations it is important to know how widespread the production of aflatoxins is, so a great deal of effort has been spent on the analysis of these metabolites. At very low concentrations a biological test is thought to be the safest confirmatory assay

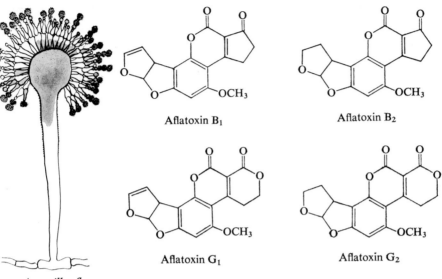

Aflatoxin B₁

Aflatoxin B₂

Aflatoxin G₁

Aflatoxin G₂

Aspergillus flavus

FIG. 6. An optical section of a sporophore of *Aspergillus flavus* and the structures of the first four aflatoxins to be described

for the production of the aflatoxins. However their intense fluorescence in ultraviolet light makes it possible to detect as little as 0·004 μg on a thin layer chromatogram (Coomes *et al.*, 1965). By most careful chromatography the four original aflatoxins can be separated as a pair of blue and a pair of green fluorescent spots with characteristic R_f values. The identification of fluorescent

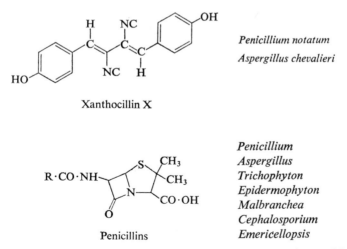

Penicillium notatum

Aspergillus chevalieri

Xanthocillin X

Penicillium
Aspergillus
Trichophyton
Epidermophyton
Malbranchea
Cephalosporium
Emericellopsis

Penicillins

FIG. 7. Two examples of unusual metabolites produced by several taxa of fungi

spots on thin layer chromatograms must involve the use of several solvent systems and authentic standards and, if possible, the preparation and chromatography of derivatives for there are a number of reports in the literature that indicate how it may be possible to misinterpret the observation of fluorescent zones on chromatoplates. Compounds with green fluorescence like aflatoxin G are reported to be produced by Japanese industrial moulds (Sasaki et al., 1968). These compounds include such metabolites as deoxymuta-aspergillic acid and 2-hydroxy-6-(1-hydroxy-1-methylpropyl)-3-sec-butylpyrazine. The presence of compounds in oats that could be mistaken for aflatoxins has also been demonstrated by Shotwell et al. (1968) and it has further been shown that the mould Macrophomena phaseoli, isolated from groundnuts, may produce a metabolite which could be confused with aflatoxin B_1 (Crowther, 1968; Schroeder, 1968).

The aflatoxins are important to us because of their biological activity and it is particularly notable, though not exceptional, that different species of animal vary considerably in their sensitivity to both the acute and chronic effects of these toxins (Barnes, 1970). As an acute poison aflatoxin is primarily responsible for damage of the liver in all species, but the single dose LD_{50} varies from as little as 0·37 mg/kg in ducklings to 9·0 mg/kg in mice (Table I). As far as the farmer is concerned pigs and cattle are both sensitive while sheep are practically unaffected when fed on meals heavily contaminated with aflatoxin; such meals may contain up to 10 ppm of aflatoxin B_1.

TABLE I

Acute oral single dose toxicities of the aflatoxins

Toxicity
Single dose LD_{50} (day old ducklings)

Aflatoxin	mg/kg
B_1	0·36
B_2	1·9
G_1	0·78
G_2	3·5

LD_{50} of aflatoxin B_1 in different animals

	(mg/kg)
Duckling	0·35–0·56
1-day-old rat	1·0
21-day-old rat	5·5
Adult rat	7·2
30-day-old hamster	10·2
Guinea pig	1·4
Weanling rabbit	0·5
Adult dog	0·5
Mouse	9·0

It is as a chronic toxin and as a liver carcinogen, that aflatoxins cause most concern and here again different species of animals respond quite differently. For the rat, which is among the least sensitive of animals as far as acute toxicity is concerned, aflatoxin is said to be the most active liver carcinogen known (Butler and Barnes, 1968). Tumours may arise following a single sublethal dose, although it may be at some considerable time after the dose has been administered. Some animals, such as the guinea pig, although much more sensitive to the acute effects of aflatoxin, do not acquire liver tumours after ingestion of sublethal doses. The mouse is not only relatively insensitive to the acute toxicity of aflatoxin but neither does it respond to these metabolites as carcinogens. It is not known how man behaves after consuming aflatoxin but there are areas of the world in which liver cancer in man occurs at a relatively high level and studies have not yet ruled out the possibility that aflatoxin, or other mycotoxins, may be implicated in the aetiology of this disease (Oettle, 1965). The fact that some animals show an increased sensitivity to the carcinogenic effects of aflatoxin when fed on protein deficient diets (McLean and McLean, 1969) is added cause for concern about those areas of the world where malnutrition is rife.

It has become clear from reports such as those of Wilson *et al.* (1967) that many diseases of previously unknown aetiology are in fact due, in part at least, to the presence of aflatoxin in food. A mysterious liver disease of pet dogs, canine X disease, is now known to be due to the presence of aflatoxin in dog foods. A dramatic increase in the occurrence of liver tumours in rainbow trout was very mystifying, until it was demonstrated that the dry feeds used in trout hatchery ponds contained aflatoxin (Halver, 1965). So sensitive are young rainbow trout to the carcinogenic activity of aflatoxin that no-one has set a minimum dose, exposure to which does not cause hepatomas. Neither is the biological activity of the aflatoxins restricted to the animal kingdom. They have been shown by Schoental and White (1965) to induce albinism in young plant seedlings and several workers indicate that aflatoxins may indeed play a part in a seedling disease of the groundnut caused by *A. flavus* (El-Khadem, 1968; Chohan and Gupta, 1968). A limited range of bacteria, mostly of the genus *Bacillus*, are inhibited by the presence of about 30 μg/ml of aflatoxin (Burmeister and Hesseltine, 1966) and *B. megaterium* has been suggested as suitable for a rapid confirmatory test for aflatoxin B_1 (Clements, 1968).

Although all the evidence available indicates that the aflatoxin molecules are themselves the toxic and carcinogenic agents, and do not have to be altered in any way by the metabolic processes of animal tissues before toxicity is expressed, some animal tissues are able to hydroxylate aflatoxins with the production of a new group of compounds referred to as aflatoxin M (Iongh *et al.*, 1964). This group of compounds, which are as acutely toxic as the parent compounds, first caused concern when they were found excreted in the milk of cows fed on aflatoxin-containing feeds, hence their designation as M compounds. However, experiments carried out at Weybridge and Carshalton in which rats

134   M. O. MOSS

were exposed for two years to a diet containing 50% of dried milk obtained from cows which themselves were fed on a diet containing 1·5 ppm of aflatoxin, gave negative results. It was concluded that, even in the event of contaminated feeds being given to dairy cows, milk would probably not offer a significant hazard, in the United Kingdom at least (Barnes, 1970).

It was soon recognized that aflatoxin M is a natural metabolite of some strains of *A. flavus* and *A. parasiticus* and it was from these sources that sufficient material was first obtained for the elucidation of structure (Holzapfel *et al.*, 1966). Four hydroxylated aflatoxins (Fig. 8) have now been isolated from strains of *Aspergillus* although it should be mentioned that addition of water to the vinyl ether double bond to give the aflatoxins B_{2A} and G_{2A} described by

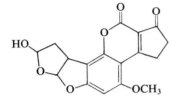

Aflatoxin B_{2a}

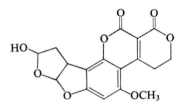

Aflatoxin G_{2a}

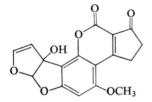

Aflatoxin M_1

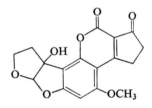

Aflatoxin M_2

Fig. 8. The four hydroxylated aflatoxins isolated from *Aspergillus flavus*

Dutton and Heathcote (1966) may occur spontaneously at low pH values. The fused difuran ring system of the aflatoxins is an intriguing feature both chemically and biochemically and it is worth considering how widespread such a structural feature is.

From *A. flavus* and *A. parasiticus* themselves, at least eleven metabolites have been isolated containing the fused difuran ring system. As well as the four original aflatoxins, and the four hydroxylated forms just referred to, a metabolite called, either parasiticol by Stubblefield *et al.* (1970), or aflatoxin B_3 by Heathcote and Dutton (1969), has been reported (Fig. 9). This compound can be thought of as aflatoxin G_1 in which the terminal δ-lactone has been hydrolysed and the resulting acid decarboxylated. It is toxic but probably not as toxic as the previously known pentacyclic aflatoxins. All these compounds so far described contain the coumarin nucleus, but in 1968 two more compounds

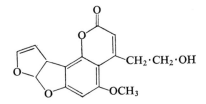

Parasiticol (Aflatoxin B₃)

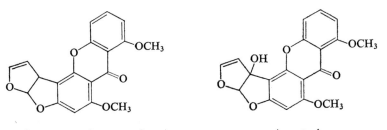

O-Methylsterigmatocystin   Aspertoxin

FIG. 9. Metabolites of *Aspergillus flavus* containing the fused difuran ring system

were isolated from *A. flavus* in which the fused difuran ring system is attached to a xanthone nucleus. These two compounds, O-methylsterigmatocystin described by Burkhardt and Forgacs (1968) and aspertoxin described by Rodricks *et al.* (1968), are derivatives of the metabolite sterigmatocystin first

Metabolites of *Aspergillus versicolor*

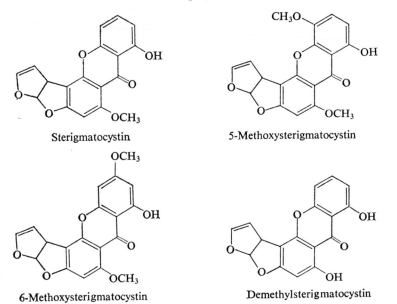

Sterigmatocystin   5-Methoxysterigmatocystin

6-Methoxysterigmatocystin   Demethylsterigmatocystin

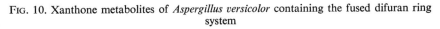

FIG. 10. Xanthone metabolites of *Aspergillus versicolor* containing the fused difuran ring system

isolated from *A. versicolor* by Hatsuda and Kuyama (1954) and described by Bullock *et al.* (1962).

A. versicolor (Vuill) Tiraboschi has been as rich a source of compounds containing the fused difuran ring as has *A. flavus*, but aflatoxins themselves have not, to my knowledge, been isolated from this source. The difuran derivatives from *A. versicolor* fall into two groups; those based on the xanthone nucleus such as sterigmatocystin itself, its 5- and 6-methoxy derivatives described by Holker and Kagal (1968) and Bullock *et al.* (1963) respectively and demethylsterigmatocystin described by Elsworthy *et al.* (1970) (see Fig. 10). Another group are the anthraquinones such as the versicolorins described by Hamasaki *et al.* (1965) and, more recently, 6-deoxyversicolorin A described by Elsworthy *et al.* (1970) (see Fig. 11).

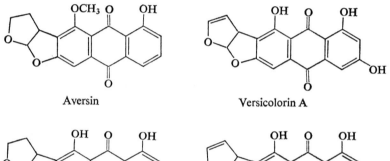

Aversin   Versicolorin A

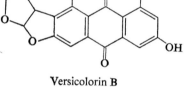   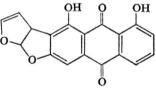

Versicolorin B   6-Deoxyversicolorin A

FIG. 11. Anthraquinone metabolites of *Aspergillus versicolor* containing the fused difuran ring system

A third extensive group of metabolites from *A. versicolor* are also anthraquinones but the branched four carbon atom system of the fused difuran rings is replaced by a linear chain of six carbon atoms. This group includes averufin and its derivatives (Holker *et al.*, 1966), averythrin (Roberts and Roffey, 1966), averantin (Roffey *et al.*, 1967) and 2-hexoyl-1,3,6,8-tetrahydroxyanthraquinone (Hamasaki *et al.*, 1967) (see Fig. 12).

The occurrence of metabolites with the fused difuran ring system was thought to be restricted to a few groups of Aspergilli, and perhaps Penicillia (the racemate of sterigmatocystin has been described as a pigment of *P. luteum* by Dean, 1963) until 1970. At that time an anthraquinone with a fused difuran ring system was reported to be produced by an organism taxonomically far removed from *Aspergillus. Dothistroma pini* is a member of the Sphaeropsidales responsible for a world wide needle blight of pines. Dothistromin, a metabolite characterized from this organism by Bear and his colleagues (1970), has a striking resemblance to some of the anthraquinone metabolites of *A. versicolor* (Fig. 13).

A consideration of all these metabolites is of interest because there has been for some time a suggestion that they may all be involved in a common bio-synthetic pathway: anthraquinones → xanthones → coumarins. Two variations on this scheme had been suggested on theoretical grounds by Holker and

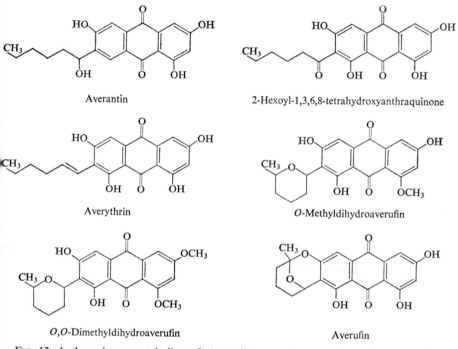

Averantin

2-Hexoyl-1,3,6,8-tetrahydroxyanthraquinone

Averythrin

O-Methyldihydroaverufin

O,O-Dimethyldihydroaverufin

Averufin

FIG. 12. Anthraquinone metabolites of *Aspergillus versicolor* containing a linear C_6 side chain

Underwood (1964) and by Thomas (1965) but only the scheme set out by Thomas (Fig. 14) could account for the results of the elegant labelling studies of Büchi and his colleagues at Massachusetts Institute of Technology (Biollaz *et al.*, 1968). Although labelled sterigmatocystin could not be incorporated into the aflatoxins (see Holker and Underwood, 1964) it has been demonstrated by Elsworthy *et al.* (1970) that 5-hydroxydihydrosterigmatocystin, labelled with ^{14}C in the *O*-methyl group, can be incorporated into aflatoxins B_2 and G_2 by *A. parasiticus*. Although the label is a potentially exchangeable group, incorporation of the whole molecule was indicated by the observation

6*

that aflatoxin B_2 and G_2 were very much more heavily labelled than were aflatoxins B_1 and G_1. Biollaz *et al.* (1968) demonstrated that labelled acetate was incorporated into the aflatoxin molecule and degradation studies on such labelled toxin indicate that all the metabolites discussed during the last few paragraphs may be produced from a single group of precursors, themselves synthesized by a head to tail linkage of nine or ten acetate units (see Fig. 15). The endoperoxide rearrangement suggested by Biollaz *et al.* (1968) as a mechanism for generating the fused difuran ring system is admittedly without a precedent amongst known chemical rearrangements but it is an elegant

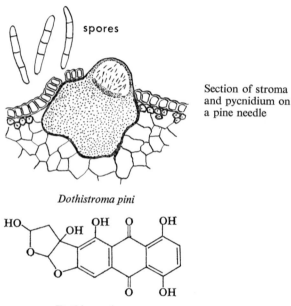

spores

Section of stroma
and pycnidium on
a pine needle

Dothistroma pini

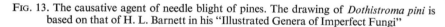

Dothistromin

FIG. 13. The causative agent of needle blight of pines. The drawing of *Dothistroma pini* is based on that of H. L. Barnett in his "Illustrated Genera of Imperfect Fungi"

hypothesis. An alternative unpublished hypothesis has been proposed by Dr. R. Thomas (personal communication) based on the possibility that the linear C_6 chain of the averufin type of anthraquinone may be a precursor of the branched C_4 chain of the versicolorins and aversin. It is suggested that the rearrangement may be initiated by the oxidative removal of a terminal acetyl group (see Fig. 17).

The polyacetate, or polyketide, route is involved in the biosynthesis of a fascinating array of metabolites by microorganisms. The diversity of what have been called secondary metabolites amongst microorganisms is comparable to the rich morphological diversity that we find in the higher animals and plants (for a discussion of secondary metabolites see, for example, Bu'Lock,

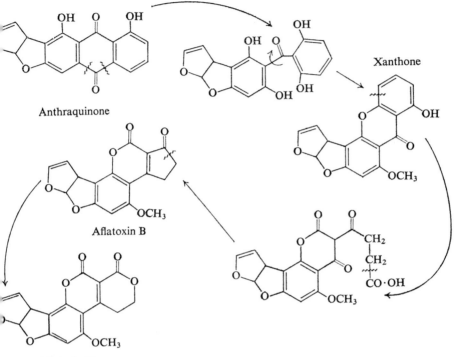

Xanthone

Anthraquinone

Aflatoxin B

Aflatoxin G

FIG. 14. A possible relationship between anthraquinones of the versicolorin type, xanthones of the sterigmatocystin type and aflatoxins (after Thomas, 1965)

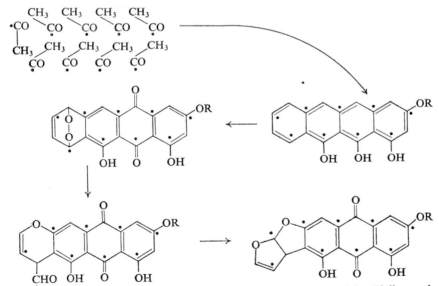

FIG. 15. A possible biogenetic origin of the fused difuran ring system (after Biollaz et al., 1968)

M. O. MOSS

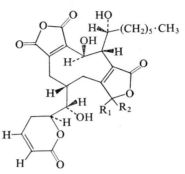

1961 and Weinberg, 1970). Because of the importance of such metabolites as the aflatoxins it is relevant to ask, what is the origin of the specificity by which the simple polyketide chain is folded, condensed, oxidized, reduced, cleaved and rearranged to give metabolites for which there often appears to be no obvious function? Although there is no doubt about the biological activity of the aflatoxins, it is not possible to be certain that this known activity is associated with an increased survival advantage for those strains which

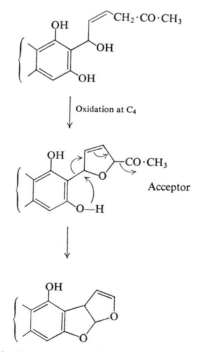

Fig. 17. A possible mechanism for deriving the difuran ring system from a linear C_6 chain (after R. Thomas, personal communication)

produce them. The suggestion has been made (Reiss, 1970) that aflatoxin production gives the producing strain a favourable advantage and it has been shown that aflatoxin can inhibit the sporulation of several fungi (see Reiss, 1971) as well as the growth of some moulds under certain cultural conditions (Lillehoj et al., 1967). However, these few reports of the behaviour of pure cultures under laboratory conditions do not conclusively demonstrate an ecological role for aflatoxin production.

III. OTHER MYCOTOXINS

Aflatoxin has justifiably excited a great deal of international and inter-disciplinary interest overshadowing the existence of many other toxic mould metabolites some of which, such as ochratoxin produced by A. ochraceus, are as acutely poisonous as aflatoxin in laboratory studies (Van der Merwe et al., 1965).

One rather general disorder amongst farm animals in which A. flavus and aflatoxin have been implicated, is the mouldy corn toxicosis of cattle and pigs reported in the veterinary literature of the U.S.A. (Burnside et al., 1957; Bailey and Groth, 1959). Although many of the symptoms could be reproduced on the assumption that aflatoxin alone was responsible, the natural disease is usually more complex and other fungi, not unexpectedly, present on moulded maize were examined for toxicity.

Penicillium rubrum Stoll was one such mould shown to be present on toxic animal feeds. It was demonstrated by Burnside et al. (1957) that this mould could produce what seemed to be a more virulent toxin than the, then unknown, toxins of A. flavus. Thus it required much smaller quantities of corn contami-nated with pure cultures of P. rubrum than of corn similarly contaminated with A. flavus to cause the death of an animal.

Two toxic metabolites, rubratoxins A and B, have been isolated and characterized from P. rubrum (see Moss et al., 1968, 1969). It can be seen from their structures (Fig. 16) that they bear little relationship to the aflatoxins and once they had been purified it was evident that they do not show the type of acute oral toxicity associated with aflatoxin. The LD_{50} of rubratoxin B when delivered in dimethylsulphoxide intraperitoneally in mice is as little as 0·3 mg/kg (Edwards and Wogan, 1968) although in propylene glycol the corresponding figure is 3·0 mg/kg. (Rose and Moss, 1970.) In the same animals the oral LD_{50} is much closer to 100 mg/kg, a factor of at least ten greater than the figure for aflatoxin B_1. The reason why P. rubrum has appeared to be so toxic is almost certainly because some strains are able to produce rubratoxins in very large quantities.

Although the rubratoxins are chemically and biosynthetically unrelated to aflatoxins, and although they are not as acutely toxic and show none of the carcinogenic properties of these latter compounds, it has been demonstrated by Edwards and Wogan (1968) that rubratoxin may potentiate the acute, but not the carcinogenic, toxicity of aflatoxin. That is to say the two compounds may act synergistically. *A. flavus* and *P. rubrum* are frequently isolated from the same sources, and indeed it is possible to demonstrate that strains of *P. rubrum* may grow parasitically upon some strains of *A. flavus* at least under laboratory conditions. It has long been known that other members of the Biverticillata symmetrica may actively parasitize such large headed species of *Aspergillus* as *A. niger*. It is not suggested that *A. flavus* and *P. rubrum* need to have such a close relationship as parasitism in the field but they certainly have similar saprophytic abilities and the combination of aflatoxins and rubratoxins may be found to account for the complex symptoms associated with mouldy corn toxicosis.

BIBLIOGRAPHY

REVIEW ARTICLES

Borker, E., Insalata, N. F., Levi, C. P. and Witzeman, J. S. (1966). Mycotoxins in feeds and foods. *Adv. appl. Microbiol.* **8**, 315.

Brook, P. J. and White, E. P. (1966). Fungus toxins affecting mammals. *Am. Rev. Phytopath.* **4**, 171.

Ciegler, A. and Lillehoj, E. B. (1968). Mycotoxins. In "Advances in Applied Microbiology" (W. W. Umbreit, ed.), Vol. 10, pp. 155.

Feuell, A. J. (1966). Toxic factors of Mould origin. *Can. Med. Assoc. J.* **94**, 1.

Goldblatt, L. A. (ed.) (1969). "Aflatoxin, Scientific Background, Control and Implications." Academic Press Inc, New York and London.

Hesseltine, C. W. (1969). Mycotoxins. *Mycopath. Mycol. Appl.* **39**, 371.

Kraybill, H. F. and Shimkin, M. B. (1964). Carcinogens related to foods contaminated by processing and fungal metabolites. *Adv. Cancer Res.* **8**, 191.

Mateles, R. I. and Wogan, G. N. (1967). Aflatoxins. *Adv. Microbial Physiol.* **1**, 25.

Moreau, Cl. (1968). Moissisures toxiques dans l'alimentation. Editions Paul le Chevalier, Paris.

Moss, M. O. (1969). Mycotoxins. *Int. Biodetn. Bull.* **5**, 141.

Pelhate, J. (1966). Dangerous moulds in animal feedstuffs. *C. r. hebd. Seanc. Acad. Fr.* **52**, 805.

Townsend, R. J. (1967). Toxic moulds and their metabolites. *Int. Biodetn. Bull.* **3**, 47.

Tyler, V. E. (1963). Poisonous mushrooms, *Progr. chem. Toxicol.* **1**, 339.

Wilson, B. J. (1968). Mycotoxins. In "The Safety of Foods," p. 141. AVI Publishing Co.

Wogan, G. N. (1969). Naturally occurring carcinogens in foods. *Progr. Exp. Tumour Res.* **11**, 134.

Wogan, G. N. and Mateles, R. I. (1968). Mycotoxins, *Progr. Indust. Microbiol.* **7**, 149.

REFERENCES

Asao, T., Büchi, G., Abdel-Kader, M. M., Chang, S. B., Wick, E. L. and Wogan, G. N. (1963). *J. Am. Chem. Soc.* **85**, 1706.

Bailey, W. S. and Groth, A. H. (1959). *J. Am. Vet. Med. Ass.* **134**, 514.

Barnes, J. M. (1970). *J. appl. Bacteriol.* **33**, 285.

Bear, C. A., Waters, J. M., Waters, T. N., Gallagher, R. T. and Hodges, R. (1970). *Chem. commun.* 1705.

Biollaz, M., Büchi, G. and Milne, G. (1968). *J. Am. chem. Soc.* **90**, 5017.

Blount, W. P. (1961). *Turkeys* **9**, 52.

Büchi, G., Foulkes, D. M., Kurono, M. and Mitchell, G. F. (1966). *J. Am. chem. Soc.* **88**, 4534.

Büchi, G., Snader, K. M., White, J. D., Gougoutas, J. Z. and Singh, S. (1970). *J. Am. chem. Soc.* **92**, 6638.

Bullock, E., Roberts, J. C. and Underwood, J. G. (1962). *J. chem. Soc.* 4179.

Bullock, E., Kirkaldy, D., Roberts, J. C. and Underwood, J. G. (1963). *J. chem. Soc.* 829.

Bu'Lock, J. D. (1961). *Adv. appl. Microbiol.* **3**, 293.

Burkhardt, H. and Forgacs, J. (1968). *Tetrahedron* **24**, 717.

Burmeister, H. R. & Hesseltine, C. W. (1966). *Appl. Microbiol.* **14**, 403.

Burnside, J. E., Sippel, W. L., Forgacs, J., Carll, W. T., Atwood, M. B. and Doll, E. R. (1957). *Am. J. Vet. Res.* **18**, 817.

Butler, W. H. and Barnes, J. M. (1968). *Fd. Cosmet. Toxicol.* **6**, 135.

Cheung, K. K. and Sim, G. A. (1964). *Nature, Lond.* **201**, 1185.

Chohan, J. S. and Gupta, V. K. (1968). *Indian J. agric. Sci.* **38**, 568.

Clements, N. L. (1968). *J. Ass. Offic. Analyt. Chem.* **51**, 1192.

Cole, M. (1966). *Process Biochem.* **1**, 334.

Coomes, T. J., Crowther, P. C., Francis, B. J. and Stevens, L. (1965). *Analyst.* **90**, 492.

Coveney, R. D., Peck, H. M. and Townsend, R. J. (1966). "Recent advances in Mycotoxicoses." S.C.I. Monograph No. 23, p. 31.

Crowther, P. C. (1968). *Analyst* **93**, 623.

Dean, F. M. (1963). *In* "Naturally Occurring Oxygen Ring Compounds." Published by Butterworths, p. 526.

Dutton, M. F. and Heathcote, J. G. (1966). *Biochem. J.* **101**, 21P.

Edwards, G. S. and Wogan, G. N. (1968). *Fedn. Proc.* 552.

El-Khadem, M. (1968). *Phytopath. Z.* **61**, 218.

Elsworth, G. C., Holker, J. S. E., McKeown, J. M., Robinson, J. B. and Mulheirn, L. J. (1970). *Chem. commun.* 1069.

Halver, J. E. (1965). *In* "Mycotoxins in Foodstuffs" (G. W. Wogan, ed.). M.I.T. Press, Cambridge, Mass.

Hamasaki, T., Hatsuda, Y., Terashima, N. and Renbutsu, M. (1965). *Agric. Biol. Chem.* (*Tokyo*) **29**, 166.

Hamasaki, T., Hatsuda, Y., Terashima, N. and Rebutsu, M. (1967). *Agric. Biol. Chem.* (*Tokyo*) **31**, 11.

Hatsuda, Y. & Kuyama, S. (1954). *J, Agric. Chem. Soc. Japan* **28**, 989.

Heathcote, J. G. and Dutton, M. F. (1969). *Tetrahedron* **25**, 1497.

Hesseltine, C. W. (1965). *Mycologia* **57**, 149.

Hesseltine, C. W., Sorenson, W. G. and Smith, M. (1970). *Mycologia* **62**, 123.

Hodges, F. A., Zust, J. R., Smith, H. R., Nelson, A. A., Armbrecht, B. H. and Campbell, A. D. (1964). *Science, N.Y.* **145**, 1439.

Holker, J. S. E. and Kagal, S. A. (1968). *Chem. Commun.* 1574.

Holker, J. S. E., Kagal, S. A., Mulheim, L. J. and White, P. M. (1966). *Chem. Commun.* 911.

Holker, J. S. E. and Underwood, J. G. (1964). *Chemy Ind.* (*London*) 1865.

Holzapfel, C. W., Steyn, P. S. and Purchase, I. F. H. (1966). *Tetrahedron Letters*, 2799.

Iongh, M. de, Vles, R. O. and van Pelt, J. G. (1964). *Nature, Lond.* **202**, 466.

Lancaster, M. C., Jenkins, F. P. and Philp, J. (1961). *Nature, Lond.* **192**, 1095.

Legator, M. S., Zuffante, S. M. and Harp, A. R. (1965). *Nature, Lond.* **208**, 345.

Lillehoj, E. B., Ciegler, A. and Hall, H. H. (1967). *Experientia* **23**, 187.

McLean, A. E. M. & McLean, E. K. (1969). *Br. med. Bull.* **25**, 278.

Moss, M. O., Robinson, F. V., Wood, A. B., Paisley, H. M. and Feeney, J. (1968). *Nature, Lond.* **220**, 767.

Moss, M. O., Wood, A. B. and Robinson, F. V. (1969). *Tetrahedron Letters*, 367.

Oettle, A. G. (1965). *S. Afr. med. J* **39**, 817.

Reiss, J (1970). *Mycopathologia* **42**, 225.

Reiss, J. (1971). *Arch. Mikrobiol.* **76**, 219.

Roberts, J. C. and Roffey, P. (1966). *J. chem. Soc.* 160.

Rodricks, J. V., Henery-Logan, K. P., Campbell, A. D., Stoloff, L. and Verrett, M. J. (1968). *Nature, Lond.* **217**, 668.

Roffey, P., Sargent, M. V. and Knight, J. A. (1967). *J. chem. Soc.* 2328.

Rose, H. M. and Moss, M. O. (1970). *Biochem. Pharmacol.* **19**, 612.

Rothe, W. (1954). *Dtsch. Med. Wochsch.* **79**, 1080.

Sargeant, K., Sheridan, A., O'Kelly, J. and Carnaghan, R. B. A. (1961). *Nature, Lond.* **192**, 1096.

Sasaki, Masaoki, Asao, Y. and Yokotsuka, T. (1968). *J. Agr. chem. Soc. Japan,* **42**, 288.

Schoental, R. and White, A. F. (1965). *Nature, Lond.* **205**, 57.

Schroeder, H. W. (1968). *Appl. Microbiol.* **16**, 946.

Scott, P. M., Walbeek, W. Van and Forgacs, J. (1967). *Appl. Microbiol.* **15**, 945.

Shotwell, O. L., Shannon, G. M., Goulden, M. L., Milburn, M. S. and Hall, H. H. (1968). *Cereal Chem.* **45**, 236.

Stubblefield, R. D., Shotwell, O. L., Shannon, G. M., Weisleder, D. and Rohwedder, W. K. (1970). *Agric. Fd Chem.* **18**, 391.

Thomas, R. (1965). *In* "Biogenesis of Antibiotic Substances" (Z. Vaněk and Z. Hoštálek, eds), p. 155. Academic Press, New York and London.

Tropical Products Institute Report (1967). Aflatoxin Bibliography 1960–1967. Ministry of Overseas Development.

Van der Merwe, K. J., Steyn, P. S. and Fourie, L. (1965). *J. chem. Soc.* 7083.

Van Walbeek, W., Scott, P. M. and Thatcher, F. S. (1968). *Can. J. Microbiol.* **14**, 131.

Van der Zijden, A. S. M., Blanche Koelensmid, W. A. A., Boldingh, J., Barrett, C. B., Ord, W. O. and Philp, J. (1962). *Nature, Lond.* **195**, 1060.

Weinberg, E. D. (1970). *Adv. microbial Physiol.* **4**, 1.

Wilson, B. J., Teer, A. P., Barney, G. H. and Blood, F. R. (1967). *Am. J. Vet. Res.* **28**, 1217.

CHAPTER 9

Selenium Toxicity

ALEX SHRIFT

Department of Biological Sciences, State University of New York, Binghamton, New York, U.S.A.

I. INTRODUCTION

Selenium has been recognized as a natural toxicant since the 1930s when seleniferous vegetation was discovered to be widespread in North America. Subsequently, toxic or potentially toxic seleniferous plants have been found in many other parts of the world.

The mechanism most frequently suggested for toxicity has been that selenium interferes with sulfur metabolism. The original basis for this view stems from the resemblances between sulfur and selenium compounds. Sulfur and selenium, together with oxygen and tellurium, fall in family VI of the periodic table. For most sulfur compounds of biological importance there exist selenium analogues with similar physical and chemical properties. The following pairs, for example, may be recognized: sulfate/selenate; cysteine/selenocysteine; methionine/selenomethionine; cystathionine/selenocystathionine; (one exception is sulfite/selenite). The resemblances between sulfur and selenium analogues were used by early investigators to explain their results as well as to guide the design of their experiments, and several lines of evidence developed indicating that enzymes which catalyse transformations of sulfur metabolites are inhibited by selenium compounds in typical antimetabolite fashion. It is only recently that differences in properties between the two classes of compounds have become an important consideration in evaluating the biological effects of selenium compounds.

Though an interference with sulfur metabolism is still generally accepted, a precise mechanism has never been established. The purpose of this review is to examine past evidence and offer more recent evidence which necessitates a reassessment of the accepted concept.

II. SULFUR-SELENIUM INTERRELATIONSHIPS

A. SULFUR AND SELENIUM LEVELS IN PLANTS

Hurd-Karrer, the first to carry out extensive physiological studies on selenium toxicity in higher plants, established several important relationships. In her analysis of total sulfur and total selenium in roots, stems, and leaves of a

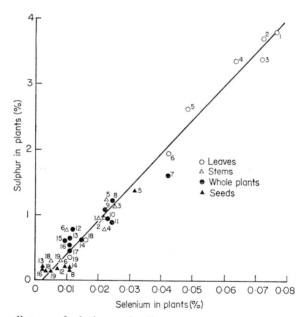

FIG. 1. Scatter diagram of selenium and sulfur percentages in leaves, stems, and seeds of various 3-month-old crop plants. See Hurd-Karrer (1937a) for plants corresponding to numbers

variety of crop plants grown with selenate, she observed that low and high sulfur contents were accompanied by low and high selenium contents, respectively (Fig. 1). Furthermore, crop plants which were normally high in sulfur, such as the Cruciferae, when grown with selenate, also absorbed considerably more selenium than did plants normally low in sulfur (Hurd-Karrer, 1935, 1936a, 1937a).

A strict parallel between sulfur and selenium does not, however, always exist. When the sulfur and selenium contents of plants from seleniferous areas were examined, the ratio of S to Se varied in different plant parts; nevertheless, there was a tendency for selenium to follow sulfur in seleniferous grains and in protein fractions from these grains (Painter and Franke, 1940).

B. COMPETITION BETWEEN SULFUR AND SELENIUM ANALOGUES

In her work with crop plants, Hurd-Karrer discovered a second important relationship between sulfur and selenium; sulfate, when supplied at appropriate levels, reduced selenate uptake and completely counteracted the toxic

Selenium (in ppm) as selenate	Degree of injury to plants grown with the following concentrations of sulfate-sulfur (in ppm)				
	0	10	32	96	192
0	0	0	0	0	0
0·01	0				
0·03	0				
0·05	0				
0·1	+				
0·2	+++	0			
0·4	++++	0			
0·6	+++++	0			
0·8	+++++	0			
1·0	Dead	+	0		
2		+++	0		
3		++++	+		
4		Dead	++		
5		Dead	+++		
6			++++		
8			++++	0	
9				0	
10				+	
11				++	
12			+++++	++	0
14				+++	0
16			Dead		0
18					+
20				++++	++
24					+++
30				+++++	++++
48				Dead	+++++
96					Dead

FIG. 2. Toxicity of selenate to wheat seedlings in relation to the sulfate concentration in the nutrient solution (injury indicated by plus signs). (Hurd-Karrer, 1936a)

effects of selenate to wheat seedlings (Hurd-Karrer, 1934, 1936a,b, 1937b, 1938). Historically, her work represents the first demonstration of a competitive antagonism between the two anions. Figure 2, taken from one of her early papers, typifies the results obtained in such competition studies. No matter what the toxic level of selenate, sulfate was able to improve growth and, at certain ratios of sulfur to selenium, eliminate toxic symptoms. Competition between sulfate and selenate has been corroborated many times by a number of investigators with higher plants, algae, bacteria, and fungi (see review by Shrift, 1967). Other pairs of analogues, such as methionine and selenomethionine, also act as competitive antagonists (Benko *et al.*, 1967; Brotherton, 1967; Cowie and Cohen, 1957; McConnell and Cho, 1965; Shrift, 1954).

The competitive nature of S *vs* Se antagonism was clearly brought out in absorption studies with excised barley roots (Leggett and Epstein, 1956). A Michaelis-Menten analysis of sulfate transport under the influence of selenate

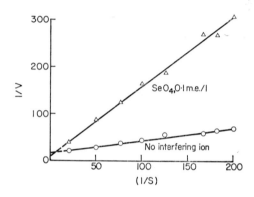

FIG. 3. Double reciprocal plot of the rate of sulfate absorption by barley roots with and without selenate (Leggett and Epstein, 1956)

gave results typical of competition for enzymes sites (Fig. 3). Selenate apparently competes for the same carrier system responsible for the active transport of sulfate.

C. REPLACEMENT OF SULFUR BY SELENIUM

The work that has been discussed can readily be interpreted to mean that selenium analogues are able to inhibit assimilation or transport or any system in which stereospecific sites for sulfur compounds are involved.

Several reports, however, are inconsistent with this interpretation. Two mutants of *Escherichia coli*, both with a nutritional requirement for methionine, were able to grow with selenomethionine instead, the remaining sulfur requirements of the cell coming from sulfate (Cowie and Cohen, 1957; Cohen and Cowie, 1957; Coch and Greene, 1971). Strain differences apparently exist, however (Wu and Wachsman, 1970). Another unusual replacement is seen with the selenium analogue of pantethine, a precursor of coenzyme A, required for growth of *Lactobacillus helveticus* (Mautner and Günther, 1959). Figure 4 shows that the organisms grew equally well with either the sulfur- or selenium-containing molecules.

Situations in which selenium replaces sulfur with little or no harmful effect are difficult to reconcile with the earlier physiological competition data.

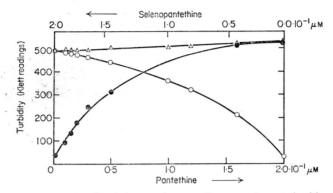

FIG. 4. Growth of *Lactobacillus helveticus* in a medium supplemented with pantethine ●——●, seleno-pantethine ○——○, or a mixture of the two △——△. (Mautner and Gunther, 1959)

III. EFFECTS OF SELENIUM COMPOUNDS ON ENZYME SYSTEMS

Data obtained with isolated enzyme systems also weaken the prevailing concept. In Table I are summarized reports of enzymes which utilize selenium analogues as substrates. These enzymes, though none is from higher plants, represent many stages in the assimilation of sulfur, from the first step to the incorporation of sulfur amino acids into protein. The first enzyme, ATP sulfurylase, converts sulfate into APS and also transforms selenate into APSe. The second enzyme in this sequence, APS kinase, apparently was unable to synthesize PAPSe (Wilson and Bandurski, 1958); the product may be unstable outside the environment of the cell. Without PAPSe it would be difficult to explain the synthesis of seleno amino acids in organisms which synthesize sulfur amino acids with PAPS as one of the intermediates.

When the kinetics of enzyme reactions with both substrates are analysed the selenium analogue often serves as well or somewhat better as substrate. The exchangeability of the two molecules is clearly seen in experiments with a methionyl-tRNA synthetase obtained from *E. coli* (Hoffman et al., 1970). The enzyme was purified on two different chromatography columns; the fractions which utilized methionine also utilized selenomethionine (Fig. 5). Analysis of the kinetic data showed that the Michaelis constants were very close in value; the affinity of the enzyme for both molecules, therefore, was the same. The results have been corroborated with a cruder preparation of the enzyme, and, in fact, the enzyme converted the selenomethionine to its tRNA derivative faster than it did methionine (Coch and Greene, 1971).

TABLE I

Replacement of sulfur substrates by selenium analogues

Enzyme	Reaction	Source	References
ATP: Sulfate-adenylyl-transferase	ATP + SeO_4^{2-} → Adenosine phosphoselenate + PP	*Saccharomyces cerevisiae*	Wilson and Bandurski (1956, 1958); Akagi and Campbell (1962)
NADPH sulfite and nitrite reductase	$SeO_3^{2-} \xrightarrow{\text{NADPH}} H_2Se$	*Escherichia coli*	Kemp *et al.* (1963)
ATP: L-methionine S-adenosyl-transferase	ATP + selenomethionine → Se-adenosylselenomethionine + PP + P	*Saccharomyces cerevisiae,* rabbit liver, rat liver	Greene (1969); Mudd and Cantoni (1957); Pan and Tarver (1967); Skupin (1962)
L-Methionine: tRNA Ligase (AMP)	ATP + selenomethionine → AMP-selenomethionine adenylate + PP	*Escherichia coli, Sarcina lutea*	Hahn and Brown (1967); Nisman and Hirsch (1958)
	ATP + selenomethionine + tRNA → L-selenomethionyl-tRNA + AMP + PP	*Escherichia coli,* rat lens	Coch and Greene (1971); Hoffman *et al.* (1970); Weller and Green (1969)
Methyltransferase	Se-adenosylselenomethionine + methyl acceptors → methylated products	Pig liver; rat liver microsomes; rat ventral prostate	Bremer and Natori (1960); Mudd and Cantoni (1957); Pegg (1969)
Acetylcholinesterase	Acetylselenocholine → cholineselenol	Rat spinal cord	Kokko *et al.* (1969)

One selenium compound that is assimilated into several seleno amino acids and also exerts a second action as an inhibitor of a variety of enzymes, though not necessarily enzymes involved in sulfur assimilation, is selenite (see reviews by Rosenfeld and Beath, 1964, and by Trelease and Beath, 1949). A strong oxidizing agent, selenite will oxidize non-biologically such compounds as cysteine and glutathione. Also it has been shown to react stoichiometrically with the eight sulfhydryl groups of reduced pancreatic ribonuclease, an inactive

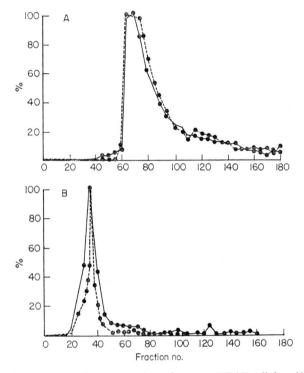

FIG. 5. Chromatography of methionyl-tRNA synthetase on DEAE-cellulose (A) and Sephadex G-200 (B); synthetase activity with methionine shown by solid line; synthetase activity with selenomethionine shown by dotted line. (Hoffman *et al.*, 1970)

form of the enzyme (Ganther, 1968; Ganther and Corcoran, 1969). The reaction of selenite with sulfhydryl groups of proteins explains the inhibition by selenite of dehydrogenases and other enzymes whose activity depends on reduced sulfhydryl groups.

It is doubtful, however, that in nature selenium toxicity involves such enzyme inhibition, for selenite does not occur in plants to any appreciable extent. Because the form of selenium in seleniferous plants consists largely of a variety of organic compounds, any mechanism of toxicity must take into account the properties of these compounds.

TABLE II.

Selenium compounds reported in plants[a]

Selenium compound	Formula
Selenium accumulators	
Se-methylselenocysteine	$CH_3-Se-CH_2-CH-COOH$ $\qquad\qquad\qquad |$ $\qquad\qquad\qquad NH_2$
Selenohomocystine	$HOOC-CH-CH_2-CH_2-Se-Se-CH_2-CH_2-CH-COOH$ $\qquad |\qquad\qquad\qquad\qquad\qquad\qquad\qquad\qquad\qquad |$ $\qquad NH_2\qquad\qquad\qquad\qquad\qquad\qquad\qquad\qquad NH_2$
γ-L-Glutamyl-Se-methylseleno-L-cysteine[b]	$HOOC-CH-CH_2-CH_2-CONH-CH-CH_2-Se-CH_3$ $\qquad |\qquad\qquad\qquad\qquad\qquad\qquad |$ $\qquad NH_2\qquad\qquad\qquad\qquad\qquad COOH$
Selenocystathionine	$HOOC-CH-CH_2-Se-CH_2-CH_2-CH-COOH$ $\qquad |\qquad\qquad\qquad\qquad\qquad\qquad\qquad |$ $\qquad NH_2\qquad\qquad\qquad\qquad\qquad\qquad NH_2$
Dimethyldiselenide	$CH_3-Se-Se-CH_3$
Nonaccumulators	
Selenocystine	$HOOC-CH-CH_2-Se-Se-CH_2-CH-COOH$ $\qquad |\qquad\qquad\qquad\qquad\qquad\qquad\qquad |$ $\qquad NH_2\qquad\qquad\qquad\qquad\qquad\qquad NH_2$

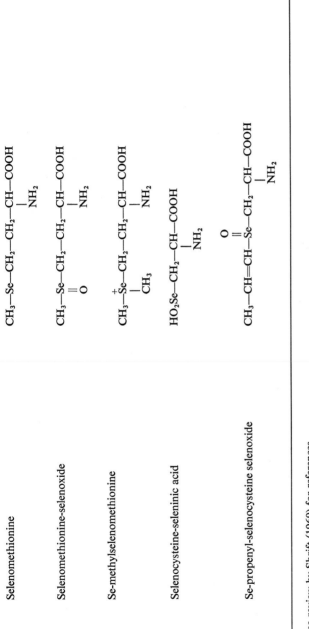

Selenomethionine

$CH_3-Se-CH_2-CH_2-CH-COOH$
                        |
                       NH_2

Selenomethionine-selenoxide

$CH_3-Se-CH_2-CH_2-CH-COOH$
        ‖ |
        O NH_2

Se-methylselenomethionine

$CH_3-\overset{+}{Se}-CH_2-CH_2-CH-COOH$
        | |
       CH_3 NH_2

Selenocysteine-seleninic acid

$HO_2Se-CH_2-CH-COOH$
                 |
                NH_2

Se-propenyl-selenocysteine selenoxide

$CH_3-CH=CH-Se-CH_2-CH-COOH$
                ‖ |
                O NH_2

[a] See review by Shrift (1969) for references.
[b] Nigam and McConnell (1969); Nigam et al. (1969).

IV. Forms of Selenium in Seleniferous Vegetation

To answer the question, why is seleniferous vegetation toxic, it becomes essential to identify the individual selenium compounds and determine their chemical and biochemical properties. In early studies, seleniferous plants showed a clear distinction on the basis of the solubility of their selenium. In seleniferous vetches, for example, most of the selenium was readily soluble in aqueous solvents (Rosenfeld and Beath, 1964; Trelease and Beath, 1949); in seleniferous grains and grasses, by contrast, most of the selenium was difficult to extract and was apparently associated with protein fractions (Painter, 1941).

The distinction between the two types of plants also is seen at the biochemical level. In seleniferous vetches the typical selenium compounds are Se-methylselenocysteine, selenocystathionine and γ-glutamyl-Se-methylseleno-cysteine (Table II). Though the presence of these compounds is presumptive evidence that they may be responsible for the acute and chronic ("blind staggers") forms of selenium poisoning in livestock, and that selenocystathionine may be responsible for several cases of human poisoning, little biochemical work, beyond their isolation and identification, has been carried out with these compounds to ascertain their fate in animals.

Much more research has been carried out with selenomethionine and selenocysteine, the compounds found in grasses and grains (Table II). On the basis of this work certain predictions, as discussed below, can be made about the mechanism of selenium toxicity.

V. Properties of Selenocysteine and Selenomethionine

Selenocysteine. This amino acid, reported in the proteins of a variety of organisms, though the evidence for its occurrence is at times equivocal (Shrift, 1967), differs in some ways from its cysteine analogue. A detailed comparison of the two amino acids is given in Table III. The pK of the —SeH is lower than that of the —SH in cysteine. At physiological pH's in a cell, diselenide bridges might therefore be favored, whereas the reduced —SH might be favored in proteins at these same pH's. Seleno groups also are more reactive with halogenated compounds, so that they would tend to oxidize more readily than would sulfhydryl groups. One could predict that replacement of cysteine by selenocysteine in proteins might alter the chemical and physical properties of proteins insofar as sulfhydryl groups and disulfide bridges were required for normal function (Huber and Criddle, 1967a).

Selenomethionine. Now widely recognized to occur in proteins and to substitute for methionine as substrate for certain enzymes, this amino acid has several properties different from those of the sulfur analogue. The infrared spectra differ. The reaction with cyanogen bromide to form homoserine is much more rapid for selenomethionine than for methionine. Significance has been attached to the lower solubility of selenomethionine in demineralized water;

TABLE III

Chemical properties of selenocysteine and cysteine[a]

Property	Selenocysteine	Cysteine
pK	5·2	8–9
Reaction rate constant ($\times 10^{-3}$ min^{-1})		
iodoacetate	2·2	1·0
iodoacetamide	0·96	0·61
chloroacetamide	0·28	0·20
Solubility	0·00235 M	0·00038 M
	(Se-cystine)	(cystine)
Fraction left after 6 h hydrolysis in	0·05	0·95
6 N HCl 110°		
Apparent half-wave potentials	−0·212 V	0·021 V

[a] Huber and Criddle (1967a).

it has been suggested that the hydrophobic properties of a protein would be affected by the replacement of methionine with selenomethionine (Shepherd and Huber, 1969).

VI. SELENO-PROTEINS

The assimilation of selenium into non-protein seleno-amino acids is thought to represent a detoxification mechanism that enables selenium accumulator plants to withstand concentrations of selenium which are toxic to other plants (Peterson and Butler, 1967; Virupaksha and Shrift, 1965). The absence of selenium from the proteins of *Neptunia amplexicaulis*, an Australian accumulator (Peterson and Butler, 1962), has been taken as an indication of such a shunt mechanism, whereas other plants which are sensitive to low levels of selenium and which assimilate selenium into proteins, lack this shunt.

From these observations a hypothesis can be developed to the effect that selenium is toxic because it is assimilated into proteins, replaces sulfur at critical places in the polypeptide chain, and gives rise to altered, malfunctioning enzymes. A few seleno-substituted polypeptides have been investigated and provide some evidence for this hypothesis.

Putidaredoxin. This protein, derived from *Pseudomonas putida*, is an iron-sulfur enzyme that hydroxylates the methylene carbon five of (+)-camphor. The protein is unusual; its sulfur is not in the polypeptide chain as an amino acid; it has two atoms of iron and two atoms of acid-labile sulfur which can be replaced by two atoms of selenium. Electron spin resonance spectra of the two types of protein differ; the two selenium atoms perhaps do not occupy precisely the same positions in the protein as do the two sulfur atoms. Nevertheless, enzyme activity is retained at levels comparable to the sulfide enzyme, though it varies between 60–100% of the activity of the sulfide enzyme. Similar results were obtained with adrenodoxin, another iron-sulfide enzyme from pig and beef adrenals (Orme-Johnson et al., 1968; Tsibris et al., 1968).

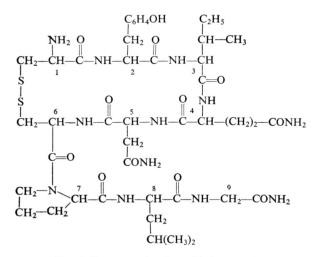

FIG. 6. Sequence of amino acids in oxytocin

Oxytocin. Figure 6 shows the structure of oxytocin, an animal hormone that elicits a variety of physiological responses. This small polypeptide has only nine amino acid residues; the two cysteine residues in the first and sixth positions of the polypeptide chain are joined by a disulfide bridge. Selenium has been substituted in the first, sixth or both positions (Walter and du Vigneaud, 1965, 1966). Table IV lists some of the physiological responses obtained with oxytocin and several of its derivatives. Some of the tests with the selenium analogues showed a decreased activity, others no significant change, and

TABLE IV

Biological potencies of oxytocin and its seleno-analogues

Compound	Depressor (fowl)	Oxytocic (rat)	Milk-ejecting (rabbit)
Oxytocin[a]	507 ± 15	546 ± 18	410 ± 16
1-Seleno-oxytocin[a]	361 ± 18	362 ± 9	351 ± 15
6-Seleno-oxytocin[a]	385 ± 15	405 ± 5	398 ± 6
Deamino-oxytocin[b]	975 ± 24	803 ± 36	
Deamino-1-seleno-oxytocin[b]	1306 ± 46	1217 ± 13	
Deamino-6-seleno-oxytocin[b]	586 ± 36	443 ± 7.0	

[a] Walter and Chan (1967).
[b] Walter et al. (1968).

several actually gave an increased response. Because this polypeptide is the only one in which the sulfur of cysteine has been partially or completely replaced by selenium, it is difficult to generalize about other proteins. Nevertheless, the results do suggest that replacement of sulfur by selenium in a disulfide bridge might impair the biological activity of a polypeptide; therefore it is conceivable that such substitutions on a wide scale within a cell would cause the cell to malfunction.

β-galactosidase. Striking results have been obtained by replacing sulfur with selenium in the enzyme β-galactosidase from *E. coli.* This large protein, with a molecular weight of about 540,000, consists of four subunits, 150 methionine residues, and about 80 cysteine residues. Workers in two laboratories have succeeded in replacing substantial amounts of the methionine with selenomethionine.

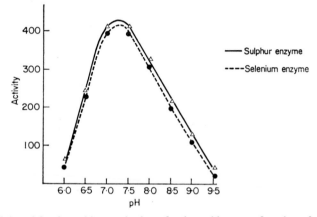

FIG. 7. Activity of β-galactosidase and seleno-β-galactosidase as a function of pH. Activity is expressed as units per g atom of protein. (Huber and Criddle, 1967b)

In one case, the seleno-enzyme was obtained from cells adapted to selenate. Fifty percent of the methionine (but apparently none of the cysteine) had been replaced by selenomethionine in the purified enzyme (Huber and Criddle, 1967b). In the second, the enzyme was isolated from another strain of *E. coli* grown with selenomethionine in the presence of cysteine; 75% of the methionine in the purified enzyme was replaced by selenomethionine (Coch and Greene, 1971). Both seleno-enzymes were as active as their sulfur counterparts. Figure 7 illustrates that the normal enzyme and the seleno-protein in which 50% of the methionine had been replaced were identical in their activities as a function of pH, with a pH optimum of 6·6; the seleno-protein with a 75% replacement had a pH optimum of 6·5.

It appears possible, therefore, to replace considerable methionine in this particular protein with no effect on its catalytic activity. In these proteins, there is probably a random replacement of the methionine residues.

Another important outcome of these studies was evidence that substitution by selenium altered the stability of the enzyme (Table V). Both heat and urea denatured the 50%-seleno-enzyme somewhat more effectively than they did the sulfur enzyme. Renaturation of the inactive seleno-enzyme was also more readily accomplished, and the activity of the reconstituted enzyme was higher than that of the renatured sulfur β-galactosidase.

TABLE V

Some properties of seleno-proteins

Seleno-β-galactosidase

Stability—less stable to heat and urea than the sulfur enzyme; renatured more readily than the sulfur enzyme.[a]

4-*S proteins*—high proportion in selenate-grown cells;[a] less stable than the sulfur 4-S proteins; contain about 120 moles of selenomethionine.

Amount produced—less in selenate-grown cells than β-galactosidase in sulfate-grown cells.[a]

Rate of induction—slower in selenomethionine-grown cells than induction of the sulfur enzyme in methionine-grown cells.[b]

Seleno-oxytocins

Configuration—energy barrier to rotation of the C—S—Se—C or C—Se—Se—C bridges is lower than the energy barrier for the C—S—S—C bridge in oxytocin.[c,d]

Stability—less stable than oxytocin; tend to form aggregates higher in molecular weight than the aggregates formed from oxytocin.[e]

---

[a] Huber and Criddle (1967b).
[b] Coch and Greene (1971).
[c] Gordon *et al.* (1968).
[d] Urry *et al.* (1968).
[e] Walter and Chan (1967).

In addition, there is evidence that the seleno-enzyme within the cell was less stable than the sulfur-enzyme. In normal cells induced to synthesize the enzyme there were seen small 4S proteins which were thought to be the subunits of β-galactosidase. In selenate-grown cells such 4S units were also prevalent, and even though the total enzyme was less than in normal cells, the proportion of these 4S units was higher; furthermore, they were richer in selenomethionine, about 120 residues, than was the seleno-enzyme (Huber and Criddle, 1967b).

Another possible indication of seleno-enzyme instability is seen from the kinetics for induction of the β-galactosidase in *E. coli* (Coch and Greene, 1971); under normal conditions enzyme activity appeared within one hour after exposure to the inducing agent; with selenomethionine present, however, the time was markedly increased (Fig. 8). This difference could also reflect a slower rate of protein synthesis. However, once synthesized, the enzyme, when isolated, apparently functioned as well as the normal enzyme.

Instability also was seen in the seleno-oxytocins. Though aggregates develop from the normal peptide, the seleno-oxytocins tended to form aggregates of higher molecular weight. Apparently the configuration of the seleno-peptide is less subject to restrictions; a C—S—Se—C or C—Se—Se—C bridge has a smaller energy barrier to rotation than does the C—S—S—C bridge, so that the seleno-peptide can be expected to differ in spatial orientation from the sulfur peptide.

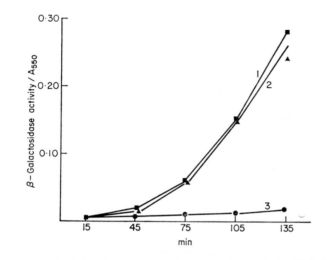

FIG. 8. Induction of β-galactosidase in E. coli 26 pregrown in enriched medium and induced in minimal medium containing (1) no further additions; (2) 4×10^{-4} M DL-methionine and 1×10^{-4} M L-cysteine; or (3) 4×10^{-4} M DL-selenomethionine and 1×10^{-4} M L-cysteine. (Coch and Greene, 1971)

VII. CONCLUSIONS

Recent biochemical evidence suggests that enzymes of many organisms utilize selenium analogues as readily as they do sulfur-containing substrates. When this occurs, selenium is assimilated into proteins where it replaces the sulfur of methionine and possibly of cysteine. As a consequence of such replacement, changes have been noted in the stability and configuration of synthetic seleno-oxytocin and of seleno-β-galactosidase isolated from E. coli. That the properties of seleno-proteins will differ from those of unsubstituted proteins can also be predicted from the properties of selenocysteine and selenomethionine.

Though stability and configuration changed, the activity of purified seleno-β-galactosidase in which 50 to 75% of the methionine had been replaced by selenomethionine remained unaltered. However, replacement of cysteine by

160   ALEX SHRIFT

selenocysteine in oxytocin caused small increases or decreases in hormonal activity, depending on the location of the replacement. The toxicity of seleniferous plants to animals and of many selenium compounds to a great variety of organisms, it seems, may rest with the properties of seleno-proteins. The following aspects of seleno-proteins bear examination: their rates of synthesis; their ability to assume the proper configuration once the polypeptide chain is completed; their catalytic activity once the proper configuration is achieved; their ability to maintain the proper configuration once synthesized; their activity in relation to the degree of substitution and to the site of substitution; and their activity in relation to substitution by selenocysteine, selenomethionine or both.

REFERENCES

Akagi, J. M. and Campbell, L. L. (1962). *J. Bact.* **84**, 1194.
Benko, P. V., Wood, T. C. and Segel, I. H. (1967). *Archs Biochem. Biophys.* **122**, 783.
Bremer, J. and Natori, Y. (1960). *Biochim. biophys. Acta* **44**, 367.
Brotherton, J. (1967). *J. gen. Microbiol.* **49**, 393.
Coch, E. H. and Greene, R. C. (1971). *Biochim. biophys. Acta* **230**, 223.
Cohen, G. N. and Cowie, D. B. (1957). *C. r. Acad. Sci. Paris*, **244**, 680.
Cowie, D. B. and Cohen, G. N. (1957). *Biochim. biophys. Acta* **26**, 252.
Ganther, H. E. (1968). *Biochemistry* **7**, 2898.
Ganther, H. E. and Corcoran, C. (1969). *Biochemistry* **8**, 2557.
Gordon, W., Havran, R. T., Schwartz, I. L. and Walter, R. (1968). *Proc. natn. Acad. Sci. U.S.A.* **60**, 1353.
Greene, R. C. (1969). *Biochemistry* **8**, 2255.
Hahn, G. A. and Brown, J. W. (1967). *Biochim. biophys. Acta* **146**, 264.
Hoffman, J. L., McConnell, K. P. and Carpenter, D. R. (1970). *Biochim. biophys. Acta* **199**, 531.
Huber, R. E. and Criddle, R. S. (1967a). *Archs Biochem. Biophys.* **122**, 164.
Huber, R. E. and Criddle, R. S. (1967b). *Biochim. biophys. Acta* **141**, 587.
Hurd-Karrer, A. M. (1934). *J. Agric. Res.* **49**, 343.
Hurd-Karrer, A. M. (1935). *J. Agric. Res.* **50**, 413.
Hurd-Karrer, A. M. (1936a). *Annual Report of the Board of Regents of the Smithsonian Institution, U.S. Govt. Printing Off., Washington, D.C. Publ. 3348*, 289–301.
Hurd-Karrer, A. M. (1936b). *J. Agric. Res.* **52**, 933.
Hurd-Karrer, A. M. (1937a). *J. Agric. Res.* **54**, 601.
Hurd-Karrer, A. M. (1937b). *Am. J. Bot.* **24**, 720.
Hurd-Karrer, A. M. (1938). *Am. J. Bot.* **25**, 666.
Kemp, J. D., Atkinson, D. E., Ehret, A. and Lazzarini, R. A. (1963). *J. biol. Chem.* **238**, 3466.
Kokko, A., Mautner, H. G. and Barrnett, R. J. (1969). *J. Histochem. Cytochem.* **17**, 625.
Leggett, J. E. and Epstein, E. (1956). *Plant Physiol.* **31**, 222.
Mautner, H. G. and Günther, W. H. (1959). *Biochim. biophys. Acta* **36**, 561.
McConnell, K. P. and Cho, G. J. (1965). *Am. J. Physiol.* **208**, 1191.
Mudd, S. H. and Cantoni, G. L. (1957). *Nature, Lond.* **180**, 1052.
Nigam, S. N., Tu Jan-I and McConnell, W. B. (1969). *Phytochemistry* **8**, 1161.
Nigam, S. N. and McConnell, W. B. (1969). *Biochim. biophys. Acta* **192**, 185.

Nisman, B. and Hirsch, M. L. (1958). *Ann. Inst. Pasteur* **95**, 615.

Orme-Johnson, W. H., Hansen, R. E., Beinert, H., Tsibris, J. C. M., Bartholomaus, R. C. and Gunsalus, I. C. (1968). *Proc. natn. Acad. Sci. U.S.A.* **60**, 368.

Painter, E. P. (1941). *Chem. Rev.* **28**, 179.

Painter, E. P. and Franke, K. W. (1940). *Am. J. Bot.* **27**, 336.

Pan, F. and Tarver, H. (1967). *Archs Biochem. Biophys.* **119**, 429.

Pegg, A. E. (1969). *Biochim. biophys. Acta* **177**, 361.

Peterson, P. J. and Butler, G. W. (1967). *Nature, Lond.* **213**, 599.

Peterson, P. J. and Butler, G. W. (1962). *Aust. J. Biol. Sci.* **15**, 126.

Rosenfeld, I. and Beath, O. A. (1964). "Selenium" Academic Press, New York and London.

Shepherd, L. and Huber, R. E. (1969). *Can. J. Biochem.* **47**, 877.

Shrift, A. (1954). *Am. J. Bot.* **41**, 345.

Shrift, A. (1967). Microbial research with selenium, *in* "Selenium in Biomedicine" (O. H. Muth, ed.). Avi Publ. Co., Westport, Conn.

Shrift, A. (1969). *A. Rev. Pl. Physiol.* **20**, 475.

Skupin, J. (1962). *Acta Biochim. Polon.* **IX**, 253.

Trelease, S. F. and Beath, O. A. (1949). "Selenium." Published by the authors, New York.

Tsibris, J. C. M., Namtvedt, M. J. and Gunsalus, I. C. (1968). *Biochem. biophys. Res. Commun.* **30**, 323.

Urry, D. W., Quadrifoglio, F., Walter, R. and Schwartz, I. L. (1968). *Proc. natn. Acad. Sci. U.S.A.* **60**, 967.

Virupaksha, T. K. and Shrift, A. (1965). *Biochim. biophys. Acta* **107**, 69.

Walter, R. and Chan, W. Y. (1967). *J. Am. chem. Soc.* **89**, 3892.

Walter, R., Gordon, W., Schwartz, I. L., Quadrifoglio, F. and Urry, D. W. (1968). *Proc. 9th European Peptide Symposium* (E. Bricas, ed.), pp. 50–55. North-Holland Publ. Co., Amsterdam.

Walter, R. and du Vigneaud, V. (1965). *J. Am. chem. Soc.* **87**, 4192.

Walter, R. and du Vigneaud, V. (1966). *J. Am. chem. Soc.* **88**, 1331.

Weller, C. A. and Green, M. (1969). *Expl Eye Res.* **8**, 84.

Wilson, L. G. and Bandurski, R. S. (1956). *Archs Biochem. Biophys.* **62**, 503.

Wilson, L. G. and Bandurski, R. S. (1958). *J. biol. Chem.* **233**, 975.

Wu, M. and Wachsman, J. T. (1970). *J. Bact.* **104**, 1393.

CHAPTER 10

Toxic Amino Acids in the Leguminosae

E. A. BELL

*Department of Botany, University of Texas at Austin,
Austin, Texas, U.S.A.*

I. Introduction 163
II. The Lathyrogens 164
  A. β-N-(γ-L-Glutamyl)aminopropionitrile 164
  B. L-α,γ-Diaminobutyric Acid 165
  C. α-Amino-β-oxalylaminopropionic Acid 165
  D. L-Homoarginine 165
  E. α-Amino-γ-oxalylaminobutyric Acid 166
  F. N-β-D-Glucopyranosyl-N-α-L-arabinosyl-α,β-diaminopropionitrile. . 166
III. Toxic Amino Acids of Other Genera 168
  A. Canavanine 168
  B. β-Cyanoalanine 169
  C. L-3,4-Dihydroxyphenylalanine 169
  D. 5-Hydroxy-L-tryptophan 170
  E. Indospicine 171
  F. Mimosine 171
  G. Azetidine-2-carboxylic Acid 172
IV. Conclusions 173
References 174

I. INTRODUCTION

Leguminous plants have proved to be a rich source of "unusual" amino acids, about sixty having been isolated from or identified in species of this family (Bell, 1971). These sixty include some that are close chemical analogues of the "usual" amino acids which are widely distributed amongst living organisms either as protein constituents or as metabolic intermediates. When such a close analogue is introduced into an organism to which it is normally foreign, then there is a very real possibility that it may act as an antimetabolite in that organism. It is not surprising therefore to find that certain of the "unusual" amino acids which have been isolated from legumes are toxic to animals (including man) insects, microorganisms and other plant species.

Several reviews on the toxicity of both natural and synthetic amino acids have appeared in recent years and for wider discussion of this subject the reader is referred particularly to those of Tschiersch (1966), Fowden et al. (1967) and Thompson et al. (1969). The present account will be restricted to those toxic amino acids which are found in species of this one plant family.

II. THE LATHYROGENS

The toxins present in the seeds of various species of Lathyrus have been the subject of research for many years because of the neurological disease (classical lathyrism) which affects men, cattle and horses when fed on diets containing a high proportion of the seed meal of L. sativus or less frequently of L. cicera and L. clymenum. In man the symptoms of classical lathyrism are loss of muscular co-ordination, paralysis of the legs and in extreme cases death. At the present time this disease is largely restricted to the sub-continent of India where outbreaks usually accompany periods of famine when alternative foods are in short supply. Accounts of the history of lathyrism have been given by Selye (1957) and Sharma (1961) while Sarma and Padmanaban (1969) discuss both the chemistry and distribution of the toxins isolated from species of this genus. In addition they list those synthetic compounds which are also, rather confusingly, referred to as lathyrogens.

One of the factors which has caused difficulty in identifying the toxin responsible for human lathyrism is the lack of biochemical uniformity exhibited by species of this genus. The genus Lathyrus may be readily divided into subgenera on the basis of the free amino acids and related compounds which accumulate in high concentration in the seeds of its various species (Bell, 1962a, 1964) and these free amino acids include at least twelve "unusual" ones.

A. β-N-(γ-L-GLUTAMYL)AMINOPROPIONITRILE

One such biochemically defined subgenus is characterized by the presence in its seeds of β-N-(γ-L-glutamyl)aminopropionitrile (1).

$$NC.CH_2.CH_2.NH.OC.CH_2.CH_2.CH(NH_2).CO_2H$$

(1) β-N-(γ-L-glutamyl)aminopropionitrile

This was the first toxic compound to be isolated from a species of Lathyrus (Schilling and Strong, 1954; Dupuy and Lee, 1954), and it was subsequently shown by Dasler (1954) that the toxicity of this compound is due to the β-aminopropionitrile (2) part of the molecule.

$$NC.CH_2.CH_2.NH_2$$

(2) β-Aminopropionitrile

This nitrile and its γ-glutamyl derivative produce skeletal abnormalities when fed to young rats (Schilling and Strong, 1954) and aortic aneurysms in both rats and chicks (Bachhuber et al., 1955).

It is now evident (Levene, 1962; O'Dell *et al.*, 1966) that the nitrile interferes with the formation of cross-linkages in collagen and elastin, possibly by inhibiting the activity of a copper containing amine oxidase (Bornstein *et al.*, 1966; Partridge *et al.*, 1964), and thereby weakens the bones and blood vessels.

B. L-α,γ-DIAMINOBUTYRIC ACID

A second subgenus of *Lathyrus* is characterized by the presence of high concentrations of L-α,γ-diaminobutyric acid (3) in its seeds. This lower homologue of ornithine was first isolated from *L. latifolius* (Ressler *et al.*, 1961) and found to be toxic to rats. O'Neal *et al.* (1968) subsequently showed that it inhibited ornithine transcarbamylase of mammalian liver.

$$H_2N.CH_2.CH_2.CH(NH_2).CO_2H$$
(3) α,γ-Diaminobutyric acid

C. α-AMINO-β-OXALYLAMINOPROPIONIC ACID

A third subgenus which contains all three of the species implicated in human lathyrism is characterized by the presence in the seeds of a strongly acidic amino acid which was identified as α-amino-β-oxalylaminopropionic acid (4) (Rao *et al.*, 1964; Murti *et al.*, 1964) and a guanidino amino acid which was identified as L-homoarginine (Bell, 1962b; Rao *et al.*, 1963).

$$HO_2C.CO.NH.CH_2CH(NH_2).CO_2H$$
(4) α-Amino-β-oxalylaminopropionic acid

α-Amino-β-oxalylaminopropionic acid was found to be neurotoxic to young birds (Adiga *et al.*, 1963), young rats, guinea pigs and dogs (Rao and Sarma, 1967). It also acts as a powerful excitant of cat spinal interneurones (Watkins *et al.*, 1966), rat brain cells (Cheema *et al.*, 1970) and produces paralysis in the hind limbs of monkeys when administered intrathecally (Rao *et al.*, 1967). The compound is not toxic when administered intraperitoneally to adult animals unless they have been previously rendered acidotic (Cheema *et al.*, 1969).

D. L-HOMOARGININE

L-Homoarginine (5) which has been found in numerous species of *Lathyrus* and also in species of *Lotus* (Chwalek and Przybylska, 1970), is not apparently toxic to mammals (Stevens and Bush, 1950; O'Neal *et al.*, 1968) but is known to inhibit growth in *Chlorella vulgaris*, in *Escherichia coli* (Walker, 1955a,b), in *Staphylococcus aureus* and in *Candida albicans* (Rao *et al.*, 1963). In the yeast *Torulopsis utilis* however homoarginine was found to reverse inhibition caused by canavanine (Walker, 1955a).

$$H_2N.C(:NH).NH.CH_2.CH_2.CH_2.CH_2.CH(NH_2).CO_2H$$
(5) Homoarginine

E. α-AMINO-γ-OXALYLAMINOBUTYRIC ACID

α-Amino-β-oxalylaminopropionic acid is accompanied in *Lathyrus* species by lower concentrations of the isomeric α-oxalylamino-β-aminopropionic acid (6).

$$H_2N.CH_2.CH(NH.CO.CO_2H).CO_2H$$
(6) α-Oxalylamino-β-aminopropionic acid

In ten of the species containing α,γ-diaminobutyric acid, both isomers of oxalyldiaminopropionic acid and the corresponding oxalyl derivatives of α-γ-diaminobutyric acid were found (Bell and O'Donovan, 1966). α-Amino-γ-oxalylaminobutyric acid (7) which occurred in higher concentrations than α-oxalylamino-γ-aminobutyric acid (8) was shown to be toxic to chicks (Rao and Sarma, 1967). The toxicity of the two α-oxalylamino acids has not been studied.

$$HO_2C.CO.NH.CH_2.CH_2.CH(NH_2).CO_2H$$
(7) α-Amino-γ-oxalylaminobutyric acid

$$H_2N.CH_2.CH_2.CH(NH.CO.CO_2H).CO_2H$$
(8) α-oxalylamino-γ-aminobutyric acid

F. N-β-D-GLUCOPYRANOSYL-N-α-L-ARABINOSYL-α,β-DIAMINOPROPIONITRILE

Low concentration of another neurotoxin tentatively identified as

(9) N-β-D-glucopyranosyl-N-α-L-arabinosyl-α,β-diaminopropionitrile

CN
.
CH.NH.Arabinosyl
.
CH.NH.glucosyl

has been isolated from seeds of *L. sativus* by Rukmini (1968, 1969) while α-amino-β-oxalylaminopropionic acid and α-oxalylamino-β-aminopropionic acid have been found in another legume genus *Crotalaria* (Bell, 1968). The distribution of known toxins in the seeds of various species of *Lathyrus* is summarized in Table I.

Other "unusual" amino acids occurring in plants of the same genus include γ-methylglutamic acid (10) (Przybylska and Strong, 1968), γ-hydroxynorvaline (11) (Fowden, 1966), γ-hydroxyhomoarginine (12) (Bell, 1963), O-oxalyl-homoserine (13) (Przybylska and Palwelkiewicz, 1965), lathyrine (14) (Bell and Foster, 1962), pipecolic acid (15) and 5-hydroxypipecolic acid (16) (Simola, 1968).

$$HO_2C.CH(CH_3).CH_2.CH(NH_2).CO_2H$$
(10) γ-Methylglutamic acid

$$CH_3.CH(OH).CH_2.CH(NH_2).CO_2H$$
(11) γ-Hydroxynorvaline

$$H_2N.C(:NH).NH.CH_2.CH_2.CH(OH).CH_2.CH(NH_2).CO_2H$$
(12) γ-Hydroxyhomoarginine

TABLE I

Toxic amino acids in seeds of *Lathyrus* species

α,γ-Diaminobutyric acid[a] — α-Amino-β-oxalylamino[a]-propionic acid — α-Amino-γ-[a] oxalylamino-butyric acid	Homoarginine[b] — Lathyrine	Lathyrine	γ-Methylglutamic acid	γ-Glutamyl-β-[a] aminopropionitrile
Lathyrus aurantius C. Koch	*L. alatus* Ten.	*L. pratensis* L.	*L. maritimus* Bigel.	*L. montanus* Bernh.
L. luteus Peterm.	*L. articulatus* L.	*L. laevigatus* sp. occidentalis Fritsch.	*L. aphaca* L.	*L. palustris* L.
L. laevigatus	*L. arvensis* Phil.	*L. varius* C. Koch	*L. alpestris* Celak	*L. aureus* Reichb. f.
sp. *aureus* Fritsch	*L. setifolius* L.	*L. niger* Bernh.	*L. variegatus* Gren & Godr.	*L. neurobolus* Boiss.
L. sylvestris L.	*L. pannonicus* Garcke	*L. machrostachys* Vogel	*L. inconspicuous* L.	*L. nissolia* L.
L. latifolius L.	*L. ochrus* DC	*L. sphaericus* Retz.	*L. incurvus* Willd.	*L. odoratus* L.
L. heterophyllus L.	*L. clymenum* L.	*L. tingitanus* L.		*L. hirsutus* L.
L. gorgoni Parl.	*L. sativus* L.[c]	*L. cyanus* C. Koch.		*L. roseus* Phil.
L. grandiflorus Sibth & Smith	*L. megallanicus* Lam.	*L. vernus* Bernh.		
L. cirrhosus Ser.	*L. quadrimarginatus* Bory & Chaub.	*L. laetiflorus* Greene.		
L. rotundifolius Willd.	*L. cicera* L.			
L. tuberosus L.				
L. multiflora Peterm.				
L. undulatus Boiss.				

[a] Toxic to higher animals.
[b] Toxic to microorganisms.
[c] Also contains N-β-D-Glucopyranosyl-N-α-L-arabinosyl-α,β-diaminopropionitrile.

$HO_2C.CO.O.CH_2.CH_2.CH(NH_2).CO_2H$

(13) O-Oxalylhomoserine

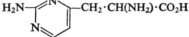

(14) Lathyrine

(15) Pipecolic acid

(16) 5-Hydroxypipecolic acid

Little is known of the effects of these compounds on other organisms, although pipecolic acid does not interfere with the utilization of proline in seedlings (Fowden et al., 1963).

III. TOXIC AMINO ACIDS OF OTHER GENERA

A. CANAVANINE

Canavanine (17) is a close analogue of arginine (18) and it occurs in many but not all genera and species of the Lotoideae (Bell, 1958, 1971; Tschiersch, 1959; Birdsong et al., 1960; Turner and Harborne, 1967). It has not been found

$H_2N.C(:NH).NH.O.CH_2.CH_2.CH(NH_2).CO_2H$

(17) Canavanine

$H_2N.C(:NH).NH.CH_2.CH_2.CH_2.CH(NH_2).CO_2H$

(18) Arginine

in species of the Mimosoideae or Caesalpinioideae (the other subfamilies of the Leguminosae) nor in species of any other plant family.

This amino acid, which can be readily identified by the characteristic magenta colour which it gives with sodium pentacyanoammonioferrate (Fearon and Bell, 1955), accumulates as the principal free amino acid in the seeds of many species. Analyses of 20 seeds of Canavalia ensiformis showed that canavanine accounted for 3·8–6·2% of the fresh weight of the embryo, while even higher concentrations (6·7–10·4%) were found in the embryos of an equal number of seeds of Dioclea megacarpa (unpublished results).

There is no shortage of evidence that canavanine is toxic to a wide range of living organisms. It inhibits growth in some but not all bacteria (Volcani and Snell, 1948; Kalyankar et al., 1958), and Richmond (1959) has demonstrated its incorporation into the protein of *Staphylococcus aureus*. It inhibits growth in moulds (Horowitz and Srb, 1948), yeasts and algae (Miller and Harrison, 1950; Walker, 1955a). It also affects growth in higher plants; Bonner (1949a,b) showed that it inhibited the development of *Avena* coleoptiles at concentrations of 3–10 mg/l whilst Steward *et al.* (1958) found that canavanine at a concentration of 10 ppm caused over 50% inhibition of growth in cultures of carrot tissue.

Walker carcinoma 256 cells (Kruse *et al.*, 1959) and rat liver ribosomal preparations (Allende and Allende, 1964) have both been shown to incorporate canavanine into protein while Morgan *et al.* (1958) found that the growth of chick heart fibroblasts was inhibited by this amino acid. The toxic effects of canavanine in whole animals as well as in tissue preparations have been shown by Tschiersch (1962). Using mice, he found that canavanine was toxic at a concentration of 200 mg/kg body weight.

These observations led Bell and Janzen (1971) to postulate that the high concentrations of canavanine which are found in seeds of *Canavalia* and *Dioclea* species may be responsible for the relative freedom from insect and animal attack which these seeds enjoy.

B. β-CYANOALANINE

β-Cyanoalanine (19) and its γ-glutamyl derivative have both been isolated from seeds of *Vicia sativa* (Ressler, 1962; Ressler *et al.*, 1963) and identified in other species of the same genus (Bell and Tirimanna, 1965). This cyanoamino acid is toxic to mammals, causing convulsions and death when injected subcutaneously into rats at a concentration of 20 mg/100 g body weight (Ressler, 1962).

The effects of β-cyanoalanine may be due, in part at least, to an inhibition of the action of vitamin B_6. This is suggested by the fact that pyridoxal hydrochloride delays and alleviates the effects of β-cyanoalanine. Rats poisoned with this compound excrete unusually large amounts of cystathione—a precursor of cystathionine on a B_6 dependent metabolic pathway (Ressler *et al.*, 1964).

$NC.CH_2.CH(NH_2).CO_2H$

(19) β-Cyanoalanine

C. L-3,4-DIHYDROXYPHENYLALANINE

L-3,4-Dihydroxyphenylalanine (L-dopa) (20) occurs in the free form or as a β-glycoside in various legume species (Guggenheim, 1913; Andrews and

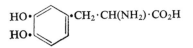

(20) L-Dopa

7*

Pridham, 1967). It was isolated in 1·5% yield from seeds of *Mucuna pruriens* by Damodaran and Ramaswamy (1937) and more recently (Bell and Janzen, 1971; Daxenbichler *et al.*, 1971) it has been found in remarkably high concentrations (6–9%) in the seed embryos of all *Mucuna* species which have been analysed. The accumulation of high concentrations of L-dopa in the seeds appears moreover to be a peculiar characteristic of this one genus.

Most studies of the toxicity of L-dopa have been related to its use as drug for the treatment of Parkinson's disease in man and the side effects seen in clinical trials include many varieties of neurological dysfunction as might be expected of this precursor of dopamine which is able, unlike dopamine itself, to cross the blood-brain barrier.

In mice, the administration of L-dopa was found to reduce the concentration of 5-hydroxytryptamine (5-HT, serotonin) in the brain and animals so treated showed irritability and a tendency to fight and jump (Everett and Borcherding, 1970).

It has been suggested (Bell and Janzen, 1971) that the role of L-dopa in the seeds of *Mucuna* species may be a protective one and that the presence of this compound, may be responsible, in part at least, for preserving these seeds from insect attack.

D. 5-HYDROXY-L-TRYPTOPHAN

5-Hydroxy-L-tryptophan (21) occurs in high concentration (6–10% dry

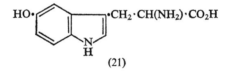

(21)

weight) in the seeds of the West African legume *Griffonia simplicifolia* and also in lower concentrations, together with 5-hydroxytryptamine (5-HT, serotonin) in the leaves (Bell and Fellows, 1966; Fellows and Bell, 1970). The plant is used in native medicine for various ailments; and is credited with being both an antiseptic and an aphrodisiac. The leaves are fed to goats in the belief that they increase fertility and are also put in hen-houses to kill lice (Irvine, 1961). Even if the plant is not quite as versatile as these reports suggest, it is a rich source of 5-HTP and some if not all of the physiological effects produced by the plant may be due to the presence of this amino acid.

Like L-dopa, 5-HTP is the precursor of a physiologically active amine in mammalian brain, in this instance 5-HT. Like L-dopa, 5-HTP can cross the blood-brain barrier and when injected into dogs and rats can produce up to tenfold increases of 5-HT in the brain (Udenfriend *et al.*, 1956, 1957). Such increases of 5-HT produce "tremors, papillary dilation, loss of light reflex, apparent blindness salivation, marked hyperpnea and tachycardia". Another effect of 5-HTP in mammals which has been reported recently is an increase in REM (rapid eye movement) sleep in man (Wyatt, 1970).

E. INDOSPICINE

The tropical legume *Indigofera spicata* (formerly *I. endecaphylla*) is hepatoxic to sheep and cows and causes abortion in the pregnant animals (Norfeldt *et al.*, 1952). It also causes severe liver damage in rabbits and mice (Hutton *et al.*, 1958).

From the seeds of this plant Hegarty and Pound (1968) isolated a new amino acid which they called indospicine and identified as L-α-amino-δ-amidinocaproic acid (22). This identification was subsequently confirmed by synthesis (Culvenor *et al.*, 1969; 1971). Purified indospicine has been shown to cause fatty

$$H_2N.C(:NH).CH_2CH_2.CH_2.CH_2.CH(NH_2).CO_2H$$

(22) Indospicine

degeneration of the liver when injected into mice and rats (Hegarty and Pound, 1968, 1970; Christie, *et al.*, 1969). This effect is reduced by the simultaneous injection of an equal weight of arginine but not by the simultaneous injection of canavanine (Hegarty and Pound, 1970).

Both extracts of the plant and the pure indospicine have been found to produce teratogenic effects (cleavage of the secondary palate) in rats (Pearn, 1967a,b; Pearn and Hegarty, 1970).

The incorporation of ^{14}C-leucine into the liver and serum protein of rats *in vivo* is depressed by indospicine (Christie *et al.*, 1969) and *in vitro* experiments have shown (Madsen *et al.*, 1970) that it inhibits the formation of ^{14}C-arginyl-tRNA (but not ^{14}C-leucyl-tRNA).

Rat liver arginase is also inhibited by indospicine (Madsen and Hegarty, 1970). The same authors showed that only the L-isomer of indospicine inhibited arginine incorporation into protein by a cell-free system.

F. MIMOSINE

Mimosine, β-[N-(3-hydroxypyridone-4)]-α-aminopropionic acid (23) was first isolated from the sap of *Mimosa pudica* by Renz (1936). Its structure was determined by Wibaut (1946) and its synthesis effected by Adams and Johnson (1949). Lysine is the precursor of this amino acid (Hylin, 1964; Tiwari and Spenser, 1965), and 3,4-dihydroxypyridine is the end product to which it is degraded by microorganisms of the soil and rumen and also by the mung bean (Smith and Fowden, 1968).

Mimosine also occurs in the seeds and leaves of various species of *Leucaena* (Brewbaker and Hylin, 1965), and it has been shown that the fresh leaves of *L. leucocephala* (*L. glauca*) may contain over 8·0% of the amino acid (Hegarty *et al.*, 1964b). Although *L. leucocephala* is extensively used in tropical agriculture, mimosine is toxic to a wide range of animals (including man) and its effects on livestock are well documented (Oakes, 1968). The most noticeable effect of mimosine on domestic animals is loss of hair, horses, donkeys, mules and pigs being particularly susceptible. More severe intoxication may however produce oedema, loss of hoof and haemorrhagic enteritis. Non-ruminants are frequently more affected than ruminants, though complete loss of wool may occur in sheep (Hegarty et al., 1964a). In rats mimosine causes a reduction in both growth rate and fertility (Hylin and Lichton, 1965). Mimosine also affects microorganisms and higher plants, inhibiting growth in both *E. coli* (Suda, 1960) and the mung bean (*Phaseolus aureus*) (Smith and Fowden, 1966).

The way in which mimosine acts is not altogether clear, and it may indeed affect different organisms differently. The inhibition of animal growth by mimosine may be prevented by adding tyrosine or phenylalanine to the diet, and it has been found that mimosine inhibits tyrosine decarboxylase from animal tissues (Crounse et al., 1962).

It has also been shown that mimosine reacts with pyridoxal phosphate and may therefore be expected to interfere with those enzymes in which pyridoxal phosphate is the prosthetic group. This is true in animals, and the amino acid inhibits enzymes such as the aspartate–glutamate transaminase of pig heart (Lin and Tung, 1964). It is not true in mung bean seedlings for even though their growth is strongly inhibited by mimosine the aspartate–glutamate trans-aminase system which they contain is not affected by the amino acid *in vitro*. Presumably, the prosthetic group is more tightly bound in the plant enzymes (Smith and Fowden, 1966). The same authors showed that the degradation product of mimosine, 3,4-dihydroxypyridine is as toxic to mung bean seedlings (but not to *Leucaena* seedlings) as is the amino acid itself.

G. AZETIDINE-2-CARBOXYLIC ACID

Azetidine-2-carboxylic acid (24) was first isolated from the leaves of *Convallaria majalis* (Lily of the Valley) by Fowden (1956). It is the principal free amino acid in the leaves of this plant and for a number of years was only known

(23) Mimosine   (24) Azetidine-2-carboxylic acid

in species of the Liliaceae and in species of the two closely related families of Agavaceae and Amaryllidaceae (Bell and Fowden, 1964). More recently, however, this imino acid has been detected in several legumes; it has been isolated from the seedlings of *Delonix regia* (Sung and Fowden, 1969), and identified chromatographically in the seeds or seedlings of species of *Bussea, Parkinsonia* and *Peltophorum* (L. Fowden, Private communication).

Azetidine-2-carboxylic acid is a lower homologue of proline and has been shown to interfere with proline synthesis or utilization in a number of different organisms. Fowden and Richmond (1963) showed that azetidine-2-carboxylic acid is activated by the proline activating enzyme of *E. coli* and *Phaseolus aureus* and incorporated into the proteins of these organisms. *Convallaria majalis* and *Polygonum multiflorum*, species which normally contain azetidine-2-carboxylic acid, do not incorporate the imino acid into their proteins and it has been shown that the proline-srRNA synthetases of *P. multiflorum* and *Ph. aureus* show different substrate specificities; the enzyme from *Ph. aureus* activates both proline and azetidine-2-carboxylic acid and that from *P. multiflorum* only proline (Peterson and Fowden, 1963; 1965). The proline activating enzyme from rat liver has also been found to activate azetidine-2-carboxylic acid (Atherly and Bell, 1964).

In *E. coli*, Baich and Smith (1968) have shown that azetidine-2-carboxylic acid interferes with proline synthesis by acting as a false end-product inhibitor. In *Saccharomyces chevalieri*, it inhibits the uptake of proline by the cells (Magana-Schwencke and Schwencke, 1969).

IV. CONCLUSIONS

It is clearly not possible to ascribe a single role to all the "unusual" amino acids which have been found in different species of the Leguminosae.

The high concentrations in which many occur, particularly in seeds, suggest that some at least have a storage function. Canavanine, for example, may account for more than 5% of a seed's weight and disappears rapidly during germination (Bell, 1960). As the percentage of nitrogen in canavanine is approximately twice that found in most storage proteins, it is tempting to suggest that it functions in the plant as a nitrogen reserve. It can be argued, moreover, that legumes which accumulate high concentrations of canavanine in their seeds have been selected because the presence of such a rich, readily available reserve of nitrogen in the seed will increase the chances of a seedling's survival (prior to the establishment of nitrogen fixation) on a nitrogen-deficient soil. Such an advantage might enable the species to colonize poor soils that would otherwise be unavailable to it. If this were the only role of canavanine in the plant, however, its presence would not confer any greater evolutionary advantage than would the presence of a similar concentration of arginine. The species containing arginine would indeed appear to be at an advantage, as all plants contain the complement of enzymes necessary for arginine synthesis whereas additional enzymes are required for the synthesis and metabolism of canavanine.

If natural selection has favoured species that store canavanine in their seeds, then canavanine must confer on them some additional advantage or advantages not provided by similar concentrations of arginine. This argument is even more compelling when we consider that some of the other "unusual"

174   E. A. BELL

amino acids which are accumulated in high concentrations in various species do not have the obvious advantages of canavanine as storage compounds.

The toxicity of many of these "unusual" amino acids to animals, insects and microorganisms suggests that some at least may protect the plants which contain them from predators. Their toxicity to plant species which do not synthesize them may also enable the synthesizing species to compete more successfully with non-synthesizing species in a particular environment. They may, thus, act as phytotoxins (see Chapter 12).

While most reports on the toxicity of "unusual" amino acids are concerned with their toxicity to man and domestic animals, it is unlikely that man or domestic animals have provided the principal environmental pressures which have led to the selection of the species of plant which contain these toxins. The pressure of insect predators might for example lead to the selection of a plant species containing an amino acid which is toxic to those particular insects and only coincidentally toxic to man. As has been seen, the same "unusual" amino acid may have very different effects in different organisms and it is clear that certain of the "unusual" amino acids, which are usually regarded as non-toxic because they are non-toxic to higher animals, may be highly toxic to other forms of life. While it may not be immediately obvious what environmental pressures have led to the selection of a species which synthesizes a particular "unusual" amino acid, the presence of that amino acid must be advantageous to the plant in terms of its overall economy or the plant would not have been able to compete successfully with other species using a higher proportion of their resources for normal vegetative purposes. It would seem more correct therefore to regard these "unusual" amino acids as aids to survival in a very competitive environment, rather than as biochemical oddities without a fundamental role in plant metabolism.

REFERENCES

Adams, R. and Johnson, J. L. (1949). *J. Am. chem. Soc.* **71**, 705–708.
Adiga, P. R., Rao, S. L. N. and Sarma, P. S. (1963). *Curr. Sci.* **32**, 153–155.
Andrews, R. S. and Pridham, J. B. (1967). *Phytochemistry* **6**, 13–18.
Allende, C. C. and Allende, J. E. (1964). *J. biol. Chem.* **239**, 1102–1106.
Atherly, A. G. and Bell, F. E. (1964). *Biochim. biophys. Acta* **80**, 510–513.
Bachhuber, T. E., Lalich, J. J., Schilling, E. D. and Strong, F. M. (1955). *Fedn Proc.* **14**, p. 175.
Baich, A. and Smith, F. I. (1968). *Experientia* **24**, p. 1107.
Bell, E. A. (1958). *Biochem. J.* **70**, 617–619.
Bell, E. A. (1960). *Biochem. J.* **75**, 618–620.
Bell, E. A. (1962a). *Biochem. J.* **83**, 225–229.
Bell, E. A. (1962b). *Biochem. J.* **85**, 91–93.
Bell, E. A. (1963). *Nature, Lond.* **199**, 70–71.
Bell, E. A. (1964). *Nature, Lond.* **203**, 378–380.
Bell, E. A. (1966). *In* "Comparative Phytochemistry" (T. Swain, ed.), pp. 159–209. Academic Press, London and New York.

Bell, E. A. (1968). *Nature, Lond.* **218**, p. 197.
Bell, E. A. (1971). *In* "Chemotaxonomy of the Leguminosae" (J. B. Harborne, D. Boulter and B. L. Turner, eds), pp. 179–204. Academic Press, London and New York.
Bell, E. A. and Foster, R. G. (1962). *Nature, Lond.* **194**, 91–92.
Bell, E. A. and Fowden, L. (1964). *In* "Taxonomic Biochemistry and Serology" (C. A. Leone, ed.), pp. 203–223. Ronald Press, New York.
Bell, E. A. and Tirimanna, A. S. L. (1965). *Biochem. J.* **97**, 104–111.
Bell, E. A. and Fellows, L. E. (1966). *Nature, Lond.* **210**, p. 529.
Bell, E. A. and O'Donovan, J. P. (1966). *Phytochemistry* **5**, 1211–1219.
Bell, E. A. and Janzen, D. H. (1971). *Nature, Lond.* **229**, 136–137.
Birdsong, B. A., Alston, R. and Turner, B. L. (1960). *Can. J. Bot.* **38**, 499–505.
Bonner, J. (1949a). *Am. J. Bot.* **36**, 325–332.
Bonner, J. (1949b). *Am. J. Bot.* **36**, 429–436.
Bornstein, P., Kang, A. H. and Piez, K. A. (1966). *Proc. natn. Acad. Sci. U.S.A.* **55**, 417–424.
Brewbaker, J. L. and Hylin, J. W. (1965). *Crop Sci.* **5**, 348–349.
Cheema, P. S., Padmanaban, G. and Sarma, P. S. (1969). *Indian J. Biochem.* **6**, 146–147.
Cheema, P. S., Padmanaban, G. and Sarma, P. S. (1970). *J. Neurochem.* **17**, 1295–1298.
Christie, G. S., Madsen, N. P. and Hegarty, M. P. (1969). *Biochem. Pharm.*, **18**, 693–700.
Chwalek, B. and Przybylska, J. (1970). *Bull. Acad. Polon. Sci.* **18**, 603–605.
Crounse, R. C., Masewell, J. D. and Blank, H. (1962). *Nature, Lond.* **194**, 694–695.
Culvenor, C. C. J., Foster, M. C. and Hegarty, M. P. (1969). *Chem. Commun.* 1091.
Culvenor, C. C. J., Foster, M. C. and Hegarty, M. P. (1971). *Aust. J. Chem.* **24**, 371–375.
Damodaran, M. and Ramaswamy, R. (1937). *Biochem. J.* **31**, 2149–2152.
Dasler, W. (1954). *Science, N.Y.* **120**, 307–308.
Dupuy, H. P. and Lee, J. G. (1954). *J. Am. Pharm. Ass.* **43**, 61–62.
Daxenbichler, M. E., Van Etten, C. H., Hallinan, E. A., Earle, F. R. and Barclay, A. S. (1971). *J. Med. Chem.* **14**, 463–465.
Everett, G. M. and Borcherding, J. W. (1970). *Science, N.Y.* **168**, 849–850.
Fearon, W. R. and Bell, E. A. (1955). *Biochem. J.* **59**, 221–224.
Fellows, L. E. and Bell, E. A. (1970). *Phytochemistry* **9**, 2389–2396.
Fowden, L. (1956). *Biochem. J.* **64**, 323–332.
Fowden, L., Neale, S. and Tristram, H. (1963). *Nature, Lond.* **199**, 35–38.
Fowden, L. (1966). *Nature, Lond.* **209**, 807–808.
Fowden, L. and Richmond, M. H. (1963). *Biochim. biophys. Acta* **71**, 459–461.
Fowden, L., Lewis, D. and Tristram, H. (1967). *Adv. Enzymol.* **29**, 89–163.
Guggenheim, M. (1913). *Z. Physiol. Chem.* **88**, 276–284.
Hegarty, M. P., Schinckel, P. G. and Court, R. D. (1964a). *Aust. J. Agric. Res.* **15**, 153–167.
Hegarty, M. P., Court, R. D. and Thorne, P. M. (1964b). *Aust. J. Agric. Res.* **15**, 168–179.
Hegarty, M. P. and Pound, A. W. (1968). *Nature, Lond.* **217**, 354–355.
Hegarty, M. P. and Pound, A. W. (1970). *Aust. J. Biol. Sci.* **23**, 831–842.
Hutton, E. M., Windrum, G. M. and Kratzing, C. C. (1958). *J. Nutr.* **65**, 321–333.
Hutton, E. M., Windrum, G. M. and Kratzing, C. C. (1958). *J. Nutr.* **65**, 429–440.
Horowitz, N. H. and Srb, A. M. (1948). *J. biol. Chem.* **174**, 371–378.
Hylin, J. W. (1964). *Phytochemistry* **3**, 161–164.
Hylin, J. W. and Lichton, I. J. (1965). *Biochem. Pharmacol.* **14**, 1167–1169.

Irvine, F. R. (1961). "Woody Plants of Ghana", pp. 308–309. Oxford Univ. Press, London.

Kalyankar, G. D., Ikawa, M. and Snell, E. E. (1958). *J. biol. Chem.* **233**, 1175–1178.

Kruse, P. F., White, P. B., Carter, H. A. and McCoy, T. A. (1959). *Cancer Res.* **19**, 122–125.

Levene, I. E. (1962). *J. Expl Med.* **116**, 119–130.

Lin, J. K. and Tung, T. C. (1964). *Kuo Li Taiwan Tah Hsueh* **10**, 69.

Magana-Schwencke, N. and Schwencke, J. (1969). *Biochim. biophys. Acta* **173**, 313–323.

Madsen, N. P., Christie, G. S. and Hegarty, M. P. (1970). *Biochem. Pharm.* **19**, 853–857.

Madsen, N. P. and Hegarty, M. P. (1970). *Biochem. Pharm.* **19**, 2391–2393.

Miller, E. J. and Harrison, J. S. (1950). *Nature, Lond.* **166**, p. 1035.

Morgan, J. F., Morton, H. J. and Pasieka, A. E. (1958). *J. biol. Chem.* **233**, 664–667.

Murti, V. V. S., Seshadri, T. R. and Venkitasubramanian, T. A. (1964). *Phytochemistry* **3**, 73–78.

Norfeldt, S., Henke, L. A., Morita, K., Matsumoto, H., Takahashi, M., Younge, O. R., Willers, E. H. and Cross, R. F. (1952). *Hawaii Agr. Exp. Sta. Tech. Bull.* 15.

Oakes, A. J. (1968). Advancing Frontiers of Plant Sciences **20**, 1–114.

O'Dell, B. L., Elsden, D. F., Thomas, J., Partridge, S. M., Smith, R. H. and Palmer, R. (1966). *Nature, Lond.* **209**, 401–402.

O'Neal, R. M., Chen, C., Reynolds, C. S., Meghal, S. K. and Koeppe, R. E. (1968). *Biochem. J.* **106**, 699–706.

Partridge, S. M., Elsden, D. F., Thomas, J., Dorfman, A., Telser, A. and Ho, P. (1964). *Biochem. J.* **93**, 30c–33c.

Pearn, J. H. (1967a). *Nature, Lond.* **215**, 980–981.

Pearn, J. H. (1967b). *Br. J. exp. Path.* **48**, 620–626.

Pearn, J. H. and Hegarty, M. P. (1970). *Br. J. exp. Path.* **51**, 34–37.

Peterson, P. J. and Fowden, L. (1963). *Nature, Lond.* **200**, 148–151.

Peterson, P. J. and Fowden, L. (1965). *Biochem. J.* **97**, 112–124.

Przybylska, J. and Palwelkiewicz, T. (1965). *Bull. Acad. Polon. Sci.* **13**, 327–329.

Przybylska, J. and Strong, F. M. (1968). *Phytochemistry* **7**, 471–475.

Rao, S. L. N., Ramachandran, L. K. and Adiga, P. R. (1963). *Biochemistry* **2**, 298–300.

Rao, S. L. N., Adiga, P. R. and Sarma, P. S. (1964). *Biochemistry* **3**, 432–436.

Rao, S. L. N. and Sarma, P. S. (1967). *Biochem. Pharmacol.* **16**, 218–219.

Rao, S. L. N., Sarma, P. S., Mani, K. S., Raghunatha Rao, T. R. and Sriramachari, S. (1967). *Nature, Lond.* **214**, 610–611.

Renz, J. (1936). *Z. Physiol. Chem.* **244**, 153–158.

Ressler, C. (1962). *J. biol. Chem.* **237**, 733–735.

Ressler, C., Redstone, P. A. and Erenberg, R. H. (1961). *Science, N.Y.* **134**, 188–190.

Ressler, C., Nigam, S. N., Giza, Y.-H. and Nelson, J. (1963). *J. Am. chem. Soc.* **85**, 3311–3312.

Ressler, C., Nelson, J. and Pfeffer, M. (1964). *Nature, Lond.* **203**, 1286–1287.

Richmond, M. H. (1959). *Biochem. J.* **73**, 261–264.

Rukmini, C. (1968). *Indian J. Biochem.* **5**, 182–184.

Rukmini, C. (1969). *Indian J. Biochem.* **7**, 1062–1063.

Sarma, P. S. and Padmanaban, G. (1969). *In* "Toxic Constituents of Plant Food Stuffs" (I. E. Liener, ed.), pp. 267–291. Academic Press, New York and London.

Selye, H. (1957). *Rev. Can. Biol.* **16**, 1–82.

Sharma, D. N. (1961). *J. Indian Med. Ass.* **36**, 299–304.

Schilling, E. D. and Strong, F. M. (1954). *J. Am. chem. Soc.* **76**, p. 2848.

Simola, L. K. (1968). *Acta Bot. Fenn.* **81**, 1–62.
Smith, I. K. and Fowden, L. (1966). *J. Expl Bot.* **17**, 750–761.
Smith, I. K. and Fowden, L. (1968). *Phytochemistry* **7**, 1065–1075.
Stevens, C. M. and Bush, J. A. (1950). *J. biol. Chem.* **183**, 139–147.
Steward, F. C., Pollard, J. K., Pachett, A. A. and Witkop, B. (1958). *Biochim. biophys. Acta* **28**, 308–317.
Suda, S. (1960). *Bot. Mag. (Tokyo)* **73**, 142–148.
Sung, M. and Fowden, L. (1969). *Phytochemistry* **8**, 2095–2096.
Thompson, J. F., Morris, C. J. and Smith, I. K. (1969). *A. Rev. Biochem.* **38**, 137–158.
Tiwari, H. P. and Spenser, I. D. (1965). *Can. J. Biochem.* **43**, 1687–1691.
Tschiersch, B. (1959). *Flora* **147**, 405–416.
Tschiersch, B. (1962). *Pharmazie* **17**, 721–730.
Tschiersch, B. (1966). *Pharmazie* **21**, 445–457.
Turner, B. L. and Harborne, J. B. (1967). *Phytochemistry* **6**, 863–866.
Udenfriend, S., Titus, E., Weissbach, H. and Peterson, R. E. (1956). *J. biol. Chem.* **219**, 335–344.
Udenfriend, S., Weissbach, H. and Bogdanski, D. F. (1957). *J. biol, Chem.* **224**, 803–810.
Volcani, B. E. and Snell, E. E. (1948). *J. biol. Chem.* **174**, 893–902.
Walker, J. B. (1955a). *J. biol. Chem.* **212**, 207–215.
Walker, J. B. (1955b). *J. biol. Chem.* **212**, 617–622.
Watkins, J. C., Curtis, D. R. and Biscoe, T. J. (1966). *Nature, Lond.* **211**, p. 637.
Wibaut, J. P. (1946). *Helv. Chim. Acta* **29**, 1669–1675.
Wyatt, R. J. (1970). *Ann. Intern. Med.* **73**, 619–622.

CHAPTER 11

Toxicity and Metabolism of *Senecio* Alkaloids

A. R. MATTOCKS

Toxicology Unit, Medical Research Council Laboratories,
Woodmansterne Road, Carshalton, Surrey, England

I. INTRODUCTION

Although the title refers to *Senecio* alkaloids, the compounds to be discussed are better described as pyrrolizidine alkaloids. The alkaloids have already been found in over a hundred *Senecio* species, and are historically associated with this genus, but they also occur in many other genera, notably in *Heliotropium*, *Lindelofia*, *Amsinckia*, and *Cynoglossum* species of the Boraginaceae, and in the Leguminosae, particularly in *Crotalaria* species. They are also found in isolated species from a number of other families.

Thus plants containing pyrrolizidine alkaloids are found in most parts of the world and the toxic species are responsible for poisoning of livestock and occasionally of humans in many countries.

The chemistry of the pyrrolizidine alkaloids has been discussed at length in reviews by Warren (1955, 1966, 1970) and by Bull *et al.* (1968). The last also covers the biological and pathological work of the Australian group very thoroughly up to 1967. The toxicology of the alkaloids has been reviewed by Schoental (1968a,b) and more recently by McLean (1970). The earlier work on toxicology and pathology is adequately covered in these reviews. Current progress in this field is rapid and in the present review I aim to cover the basic facts briefly and then to move on to some of the most recent developments. In accord with the title, the emphasis will be on results which concern the relation between metabolism of the alkaloids and their toxicity.

II. CHEMICAL STRUCTURES

Most of the alkaloids are esters of amino-alcohols such as supinidine (1), retronecine (2) and crotanecine (3), with a variety of branched chain, usually hydroxy aliphatic acids. The alkaloids themselves may be monoesters, such as heliotrine (4), diesters like lasiocarpine (5), or cyclic diesters like retrorsine (6) and monocrotaline (7). In general the first group are the least toxic and the last group the most toxic. A few alkaloids, such as senkirkine (8), are esters of the amino-alcohol otonecine, which is a hybrid molecule whose ketonic character is diminished through interaction across the ring with the amino group. This alkaloid is about one-seventh as toxic as retrorsine (6) (Schoental, 1970). Some alkaloids, like platyphylline (9), have no double bond in the pyrrolizidine ring, and these are not toxic. The alkaloids frequently occur in the plant partly or wholly in the form of their *N*-oxides. These are more water soluble than the free bases and often cannot be extracted into organic solvents. Thus, to obtain a maximum yield of alkaloids from a plant, it is usual to reduce chemically the aqueous extract to convert *N*-oxides to bases before attempting to extract the alkaloids with chloroform (Koekemoer and Warren, 1951).

It will be appreciated that there are many different alkaloids and often several will occur, together with *N*-oxides, in the same plant. The toxicity of a plant may thus be due to one or several alkaloids and it may vary with season or locality, and according to whether the plant is fresh or dried (as when harvested with hay) since some alkaloids deteriorate during storage and plant materials become less toxic on keeping (Schoental 1968b).

A field test is available for detecting the majority of toxic alkaloids in small samples of plant materials (Mattocks, 1971a), and a more sensitive analytical technique can be used for quantitative estimations of the unsaturated alkaloids and *N*-oxides (Mattocks, 1967, 1968b; Bingley, 1968).

The toxic effects of a large number of alkaloids have been studied by many workers. The structural features required for toxicity have been discussed by Schoental (1957, 1968a) and further information was obtained from the study of some semi-synthetic derivatives (Schoental and Mattocks, 1960). Thus we can define some of the requirements for toxicity in pyrrolizidine alkaloids, and these are illustrated in Fig. 1.

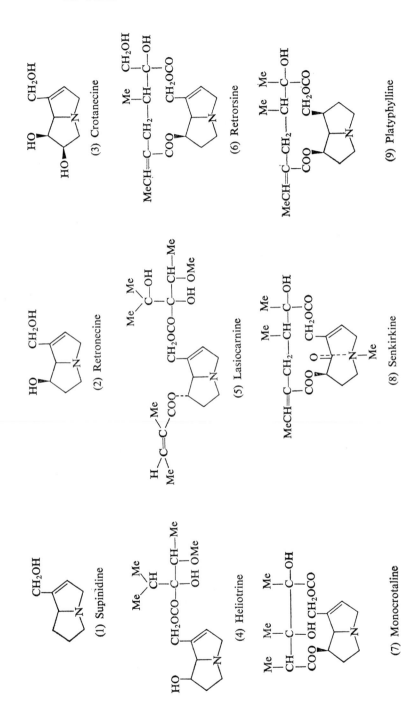

(1) Supinidine

(2) Retronecine

(3) Crotanecine

(4) Heliotrine

(5) Lasiocarnine

(6) Retrorsine

(7) Monocrotaline

(8) Senkirkine

(9) Platyphylline

The alkaloid must possess the 1-hydroxymethylpyrrolizidine system, unsaturated in the 1,2-position, and at least one of the hydroxyl groups must be esterified. This is usually the hydroxymethyl group, but slight toxicity is shown if only the 7-hydroxy group is esterified, as in 7-angelylheliotridine, and only very slight toxicity is shown if the 7-hydroxy group is absent, as in supinine (Bull *et al.*, 1968, p. 166). The greatest toxicity is shown by diesters, and especially by cyclic diesters like retrorsine (6). It was earlier thought that the acid moiety needed to possess a branched carbon chain, but recent work, discussed later in this review, shows that this is not necessarily true. The stereochemistry of the acid or basic moieties does not affect the toxicity qualitatively, but may cause quantitative differences in toxicity.

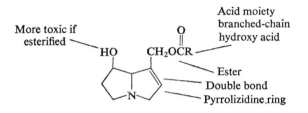

FIG. 1. Typical structure of a toxic pyrrolizidine alkaloid

III. Toxic Effects of the Alkaloids

Toxic effects of the alkaloids can be divided into peracute effects and cytotoxic effects.

Peracute effects might be seen after large doses of some of the alkaloids, and might for example affect the central nervous system, causing convulsions and death within a few minutes. We shall not be concerned with such effects here.

Cytotoxic effects are those in which cells in various organs such as liver, lungs, and perhaps others such as heart, kidneys, brain, are damaged or destroyed. This may result in death of the animal from a few days to years after receiving the alkaloid, with tissue damage which can be observed macroscopically or histologically. In this review, the term toxicity always means cytotoxicity.

Acute toxicity refers to damage to an organ which may result in death of an animal up to a week after dosing.

Chronic toxic effects are those which manifest themselves in an animal months or years after it has received a single dose of an alkaloid, or ingested repeated sublethal doses of alkaloid, as when grazing on poisonous plants.

The nature of the toxic effects caused by pyrrolizidine alkaloids may depend on such factors as the species and age of the affected animal, the structure of the alkaloid(s), and the way the alkaloid is ingested, e.g. as a single massive dose or as many small doses.

Species which can be affected include farm animals such as cattle, sheep and horses, especially in arid regions where pastures may be of poor quality. Ragwort (*Senecio jacobaea*) is well known to British farmers as a danger to livestock. People are known to have been affected, for example in Jamaica where herbal medicines or "Bush Teas" have sometimes included extracts of *Crotalaria* bushes (Bras *et al.*, 1957). Animals which have been treated experimentally with the alkaloids in the laboratory include rats, mice, hamsters, rabbits, guinea pigs, monkeys, sheep, chickens and quail, but rats have been most frequently used.

Young rats are more susceptible to the effects of the alkaloids than adults (Jago, 1970; Schoental, 1970), although we have found that this does not apply during the first few hours after birth. It has also been shown (Schoental, 1959) that alkaloids ingested by mother rats can be transmitted to suckling young in the milk, with the result that the livers of the young animals are damaged while the mother escapes injury.

Toxic effects are seen most frequently in the liver. Acute effects include massive centrilobular haemorrhagic necrosis of the liver cells (Davidson, 1935). The most characteristic chronic effect is the development within a few weeks of large numbers of giant cells (megalocytes) in the liver (Bull, 1955; Schoental and Magee, 1957a, 1959). These appear to be unable to divide, and have led to the recognition that the alkaloids are antimitotic agents (Schoental and Magee, 1959). This is seen in animals given doses of alkaloids which are too small to produce acute hepatotoxic effects, and particularly when repeated small doses are given, or when single doses are given to very young rats (Jago, 1969).

The megalocytosis may progress until almost all the liver consists of giant cells and the animal ultimately dies. The antimitotic activity of some of the alkaloids has been studied in detail by Downing and Peterson (1968), and some of them have been found to possess anti-tumor activity (McLean, 1970, p. 475).

A syndrome called veno-occlusive disease of the liver occurs in humans, particularly in the West Indies (Bras *et al.*, 1957), and is believed to be a result of acute poisoning by pyrrolizidine alkaloids. The effect has been reproduced in experimental animals (McLean, 1970, p. 464).

Some of the pyrrolizidine alkaloids have been described as liver carcinogens, especially by Schoental (1968b), and tumours have been observed in experimental animals by several workers. However, there appears to be disagreement among pathologists about what constitutes a tumour, and Bull *et al.* (1968) are not convinced by the available evidence. No large scale, properly controlled, carcinogenicity test has been carried out on any of the alkaloids, and until this is done, the issue will continue to remain in doubt.

Apart from the liver, the organs most frequently affected by pyrrolizidine alkaloids are the lungs. Lung lesions are especially common after ingestion of *Crotalaria* alkaloids such as monocrotaline and fulvine (Barnes *et al.*, 1964). Kay and Heath (1969) have published a monograph entitled "*Crotalaria spectabilis*, the Pulmonary Hypertension Plant", though this title is a little misleading since the active agent is the alkaloid monocrotaline (7), which also occurs in other plants, and it can also cause other lesions while other alkaloids can cause similar effects. A characteristic lung lesion in rats experimentally poisoned with *Crotalaria* alkaloids is the accumulation, 3–6 weeks after a single dose, of large amounts of fluid in the chest (pleural effusions). This is accompanied by some oedema within the lungs, and thickened alveolar walls containing some abnormally large cells, and death usually follows from lung failure (Barnes et al., 1964). These authors suggested that such lung damage might be caused by a toxic metabolite formed in the liver and escaping to the lungs.

Lesions in other organs have occasionally been associated with pyrrolizidine alkaloid poisoning, and references are to be found in McLean's review (1968).

IV. METABOLISM OF THE ALKALOIDS

A. REASONS FOR BELIEVING THAT TOXICITY IS DUE TO METABOLITES

There are good reasons for the view that metabolites, not the alkaloids themselves, are responsible for toxic effects.

1. Major toxic effects occur in certain organs, regardless of the route of administration of the alkaloid. Thus the liver of the rat is damaged regardless of whether the alkaloid is given by mouth or by intravenous, intraperitoneal or subcutaneous injection. Cytotoxic effects are not seen at the injection site. The alkaloids are known to be metabolized in the rat liver.

2. Direct application to the skin does not cause local toxic effects (Schoental et al., 1954).

3. The alkaloids are not toxic to some organisms, for example cinnabar moth larvae, although they may be accumulated in the body in relatively large amounts (Aplin et al., 1968).

4. The chronic toxicity of a single dose of retrorsine given to newborn rats less than 1 hour old is much less than that of a similar dose given to rats more than 1 day old. The liver microsomal metabolizing enzymes in newborn rats have very low activity towards retrorsine, but this activity increases very considerably during the first few days of life. This will be considered more fully later (Section V, E).

5. Some species of animals are more resistant to the liver damaging effects of the alkaloids than others. Thus an LD_{50} dose of retrorsine to male rats is about 40 mg/kg body weight, whereas guinea pigs sustained no liver damage at 420 mg/kg, survived doses of 800 mg/kg and died from peracute effects at 1000 mg/kg. However, if the liver microsomal metabolizing enzymes of guinea pigs are stimulated by pretreatment with phenobarbitone, the LD_{50} of retrorsine to these animals falls to about 230 mg/kg and the animals die with characteristic haemorrhagic liver necrosis.

6. The alkaloids themselves are chemically not very reactive. It is hard to envisage reactions which they could readily undergo with cell constituents under physiological conditions. Culvenor et al. (1962) suggested that they

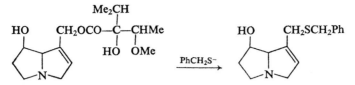

FIG. 2. Alkylation of benzyl mercaptan by the alkaloid heliotrine

might act as alkylating agents, through alkyl-oxygen fission, a known reaction of allylic esters. They showed that the rather reactive thiol, benzyl mercaptan, could be alkylated by the alkaloid heliotrine (Fig. 2), the sulphur replacing the allylic ester group on the basic moiety, but there is no evidence that a similar reaction could take place in the living cell.

7. Chemically prepared derivatives, likely to be the same or similar to metabolites of the toxic alkaloids, when administered to rats show most of the toxic effects of the parent alkaloids but at much lower dose levels. This will be further discussed later (Section VI, A).

The foregoing results may be interpreted in various ways, but taken together, they point convincingly to the likelihood that pyrrolizidine toxicity is due to metabolites.

B. POSSIBLE METABOLIC PATHWAYS

The alkaloids are known to be metabolized in a number of ways, and we shall consider whether any of the metabolites might be more toxic than the parent alkaloids.

1. Hydrolysis

The ester groups of the alkaloids are readily hydrolysed by aqueous alkali, and a comparison of chemical hydrolysis rates has been made by Bull *et al.* (1968, p. 60). In contrast we have found the natural alkaloids to be rather

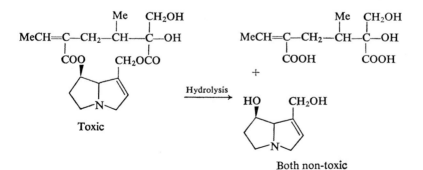

FIG. 3. Hydrolysis of a toxic pyrrolizidine alkaloid (retrorsine) to its acid and basic moieties (isatinecic acid and retronecine)

resistant to hydrolysis by esterases from rat liver or serum *in vitro* but some hydrolysis does take place in some species. Thus heliotrine and lasiocarpine undergo hydrolysis to a small extent in the rat (Bull *et al.*, 1968, p. 217) and heliotrine is hydrolysed in the sheep (Jago *et al.*, 1969). Hydrolysis products, the acid and basic moieties of some of the alkaloids, have been tested and found not to be cytotoxic (Schoental and Mattocks, 1960). Thus hydrolysis may be regarded as a detoxication process (Fig. 3).

2. N-*Oxidation*

The alkaloids are easily converted into their *N*-oxides and the reaction is readily reversed by reducing agents (Fig. 4). The oxidation occurs metabolically in the rat (Bull *et al.*, 1968, p. 317; Mattocks, 1968a,c) and in the sheep (Jago *et al.*, 1969). The *N*-oxides are highly water soluble and are rapidly excreted. Pyrrolizidine *N*-oxides are probably considerably less toxic than the parent alkaloids. According to Jago *et al.* (1969), heliotrine *N*-oxide is about one-

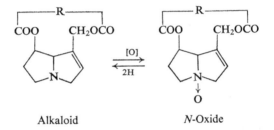

Alkaloid   *N*-Oxide

FIG. 4. *N*-Oxidation of a pyrrolizidine alkaloid

fifteenth as toxic as heliotrine to sheep. In our experiments with male rats retrorsine *N*-oxide had an LD_{50} well above 200 mg/kg by intraperitoneal injection but near to 40 mg/kg, almost the same as retrorsine alkaloid, when given by mouth. We found that the *N*-oxide was reduced to the base in the rat gut, perhaps by intestinal flora. This reducing system would not be readily accessible to *N*-oxide injected intraperitoneally or intravenously or formed metabolically in the liver (unless excreted in the bile). Thus it seems likely that the high toxicity of pyrrolizidine *N*-oxides ingested orally is due to their metabolic reduction in the gut to the alkaloid bases, which can then be absorbed and transported to the liver and that the *N*-oxides themselves are much less toxic, or non-toxic, and may be regarded as detoxication products.

3. *Detoxication in the Sheep*

Two further metabolic processes occur in the sheep (Fig. 5). Jago *et al.* (1969) showed that a methoxy group in the acid moiety of heliotrine is demethylated. The demethylated alkaloid is rather less toxic. In addition, a special detoxication reaction takes place in the rumen, whereby heliotrine is converted to two non-toxic bases (Dick *et al.*, 1963; Lanigan and Smith, 1970).

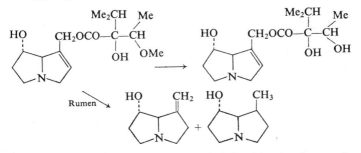

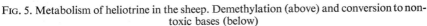

FIG. 5. Metabolism of heliotrine in the sheep. Demethylation (above) and conversion to non-toxic bases (below)

4. Epoxidation

There is no experimental evidence as yet that the double bond in the basic moieties of toxic pyrrolizidine alkaloids can be metabolically epoxidized, but the possibility is mentioned since Bull *et al.* (1968, p. 214) and, more recently, Schoental (1970) have suggested that such epoxides, being chemically reactive, might be more toxic than the parent alkaloids. However this possibility appears to have been ruled out by Culvenor *et al.* (1969a, 1971) who tested the two epoxides of monocrotaline and found them to be non-toxic. We have found that monocrotaline epoxides are not converted to pyrrolic derivatives (see below) by rat liver microsomes *in vitro*.

5. Dehydrogenation to Pyrrolic Derivatives

In the rat, toxic pyrrolizidine alkaloids are metabolized to pyrrolic derivatives, which can be detected in the urine and in tissues by the red colour which they give with a modified Ehrlich reagent (Mattocks, 1968a; Culvenor *et al.*, 1969). The reaction is brought about by the microsomal mixed-function oxidases of rat liver (Jago *et al.*, 1970; Mattocks and White, 1971b) and it amounts to dehydrogenation of the alkaloid (Fig. 6). Pyrrole derivatives are formed in the livers of a number of other species, including mice, hamsters, rabbits, quail and chickens. Unlike the original alkaloids, the dehydro-alkaloids or "pyrrole derivatives" are chemically highly reactive (Mattocks, 1968a, 1969a; Culvenor *et al.*, 1970a), and there are good reasons for believing they are responsible for some or all of the toxic effects of the alkaloids.

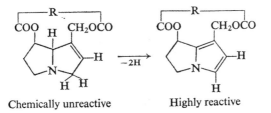

Chemically unreactive   Highly reactive

FIG. 6. Dehydrogenation of an unsaturated pyrrolizidine alkaloid to a dihydropyrrolizine ("pyrrole") derivative

V. Pyrrolic Derivatives as Toxic Metabolites

A. REACTIVITY AS ALKYLATING AGENTS

The pyrrole derivatives of the toxic alkaloids are chemically dihydro-pyrrolizine esters, in which the ester groups are "activated" by conjugation with the nitrogen. Thus an ester group may readily be lost leaving the positively charged dihydropyrrolizine moiety which can react with negatively charged (nucleophilic) groups such as amines or thiols to form relatively stable alkylation products (Fig. 7). It should be noted that in the case of diesters both ester groups can be thus activated, though not at the same time (Mattocks, 1968a). Thus the alkaloids could act as bifunctional alkylating agents. Pyrrolic derivatives have been prepared chemically from the alkaloids and such

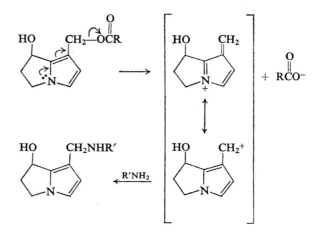

Fig. 7. Alkylation of an amine by a reactive dihydropyrrolizine ester

alkylation reaction have been carried out in the laboratory (Mattocks, 1969a; Culvenor et al., 1970a). The hydroxyl groups of non-esterified pyrrole deriva-tives can also act as alkylation sites though they are less reactive (Mattocks, 1969a). A consequence of the extreme reactivity of the pyrrolic esters is that they are hydrolysed very rapidly in water with half lives measured in seconds. Thus, when they are introduced into an aqueous environment (such as a living cell) which also contains nucleophilic groups (Y^-) there will be competing hydrolysis and alkylation reactions (Fig. 8) (Mattocks, 1969a). The hydroxy derivatives resulting from hydrolysis might also alkylate Y^-, but more slowly. However in absence of a buffer the acid released at hydrolysis will cause rapid polymerisation of the remaining pyrrolic material.

It is evident that the metabolic formation of such reactive compounds within the cell is likely to lead to alkylation of cell components, some of which might be necessary for the proper functioning of the organism, and thus cell damage or death may occur.

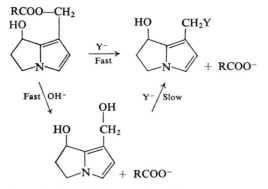

FIG. 8. Alkylation of a nucleophile (Y⁻) by a reactive dihydropyrrolizine ester in an aqueous environment. Competing alkylation and hydrolysis reactions

B. PYRROLES FROM NON-TOXIC ALKALOIDS

It has been found that the alkaloid platyphylline (9) and its close relative rosmarinine (10, Fig. 9) are both metabolized to pyrrole-like derivatives in the rat (Culvenor *et al.*, 1969a; Mattocks, 1968a). This presented a problem since these alkaloids are not cytotoxic. However investigation of the structure of the metabolite from rosmarinine (Mattocks and White, 1971a,b) has shown that it is chemically different from the pyrroles from toxic alkaloids (Fig. 9).

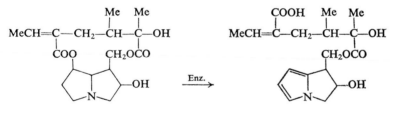

(10) Rosmarinine

FIG. 9. Conversion of rosmarinine to a non-toxic pyrrolic derivative in the rat liver

Instead, the alternative ring is dehydrogenated, with loss of one of the ester groups. The remaining ester group is not reactive because it is not conjugated with the nitrogen and so the compound is stable, cannot act as an alkylating agent, and is not cytotoxic.

C. METABOLIC FORMATION OF PYRROLE DERIVATIVES

It is easy to demonstrate that pyrrolic metabolites are formed from pyrrolizi-dine alkaloids when they are incubated aerobically with rat liver *in vitro*. Reaction with Ehrlich reagent subsequently produces red colours both in the slices and in the incubation fluid (Mattocks, 1968a). Studies using subcellular fractions of rat liver *in vitro* have shown that most of the metabolic activity is in the microsomal fraction, oxygen and reduced NADP are required, and the reaction is inhibited by carbon monoxide (Mattocks and White, 1971b). The alkaloids are also converted to *N*-oxides by this system, though not necessarily by the same enzymes (Mattocks, 1968c). However the *N*-oxides are not converted to pyrroles, and it seems likely that metabolic pyrrole formation does not proceed via the *N*-oxides (Jago *et al.*, 1970; Mattocks and White, 1971b). The rates of pyrrole formation in isolated microsome systems vary with the alkaloids, and generally are lower for the less toxic alkaloids such as heliotrine. Because of the high chemical reactivity of the primary pyrrolic ester metabolites, it is not possible to isolate these as such, but hydrolysis and alkylation products can be identified. Thus, Jago *et al.* (1970) incubated heliotrine (4) or lasiocarpine (5) with rat liver microsome preparations and isolated dehydroheliotridine (11) as the major water soluble product, together with a quaternary base (12) resulting from the alkylation of unchanged heliotrine by the reactive

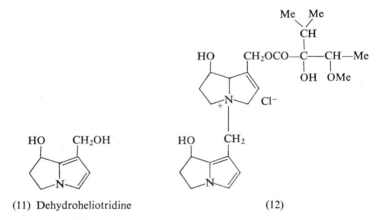

(11) Dehydroheliotridine   (12)

pyrrolic intermediate dehydroheliotrine. The same compound had earlier been found in the heliotrine-containing plant *Heliotropium europaeum* (Culvenor and Smith, 1969).

In the rat, the liver is the only organ in which the alkaloids are metabolized to pyrroles to any significant extent. Thus damage in other organs, such as the pulmonary vessels, appears to be due to metabolite which has escaped from the liver, as was earlier suggested by Davidson (1935) and by Barnes *et al.* (1964). Pyrrolic metabolites are formed in the liver of species other than rats, and have been demonstrated in human embryo liver tissue after being incubated with lasiocarpine (Armstrong and Zuckerman, 1970).

D. MECHANISM OF METABOLIC PYRROLE FORMATION

Dehydrogenation is not a common reaction of microsomal enzymes, but the so-called mixed function oxidases are well known to cause oxidative changes such as *N*-oxidation, *N*- and *C*-hydroxylation, and *N*- and *O*-dealkylations.

The general aim of such reactions appears to be to convert foreign and possibly toxic compounds into more water soluble derivatives which can be quickly eliminated from the body. The mechanism of pyrrole formation is not known, but it may be observed that hydroxylation of the pyrroline ring at a position adjacent to the nitrogen will lead to a series of chemically unstable intermediates which would decompose spontaneously to the pyrrole structure (Fig. 10) (Mattocks and White, 1971b). The otonecine esters, like senkirkine (8), could, after metabolic *N*-demethylation, undergo a similar transformation (Culvenor *et al.*, 1971; Mattocks and White, 1971b). Seen in this way, the formation of the pyrroles would appear to be the result of a metabolic "mistake", more toxic derivatives being formed where less toxic ones were intended.

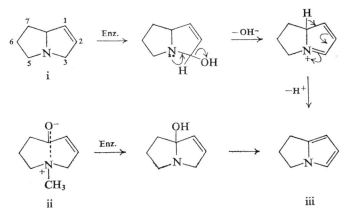

FIG. 10. Hypothetical mechanism for the enzymic conversion of an unsaturated pyrrolizidine alkaloid (i) and of an otonecine alkaloid (ii) to dihydropyrrolizine derivatives (iii). Ring substituents not directly involved in the reactions have been omitted for simplicity, but the presence of ester groups is normally necessary for dehydrogenation to take place

E. FORMATION OF PYRROLE DERIVATIVES *In vivo*

It is possible to demonstrate the presence of pyrrolic metabolites in the tissues of animals which have been dosed with toxic pyrrolizidine alkaloids (Mattocks 1968a). They are found mainly in the liver, but in smaller amounts also in the lungs, especially after ingestion of alkaloids, such as monocrotaline, which are most liable to cause lung damage. Procedures are available which can give a quantitative measure of the pyrrolic metabolites in tissue (Mattocks and White, 1970), but it is important to realize what this means. Figure 11 illustrates the hypothetical fate of a reactive pyrrolic metabolite in the liver. Some may be hydrolysed and some might alkylate soluble thiols or amines, such as glutathione or amino acids. The relatively stable, soluble products will find their way to the urine and be eliminated from the body though elimination may be slow because of adsorption on macromolecules. Some of the reactive metabolite might react with structural or other macromolecules, perhaps proteins

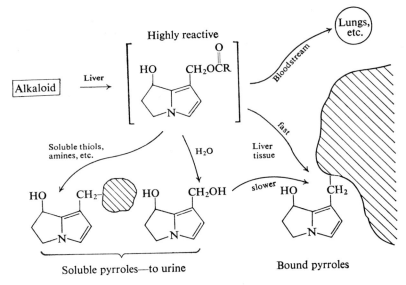

FIG. 11. Hypothetical fate of a reactive pyrrolic metabolite (dihydropyrrolizine ester) in the liver of a rat

or nucleic acids within the cell, and remain for a much longer time as "bound pyrroles". It may be supposed that some such reaction products are responsible for toxic effects in the cell. Finally, some of the reactive metabolite might survive long enough to be transported in the bloodstream to other organs. In this case reactions may take place in the blood vessels of the heart and lungs.

Thus, it will be seen that the Ehrlich colour formed in a homogenized piece of liver tissue will be due not to one metabolite, but to a series of reaction and hydrolysis products. However, experience has shown that this technique gives a good indication of the total amount of pyrrole metabolites present in the organ. Figure 12 shows the results of giving a series of male rats a constant

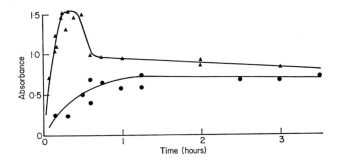

FIG. 12. Relative liver pyrrole metabolite levels in male rats at various times after giving the alkaloid retrorsine (50 mg/kg); ●, by mouth; ▲, intraperitoneally

dose (50 mg/kg) of the alkaloid retrorsine, either orally or intraperitoneally. The animals were killed after various times, and pyrrole levels were measured in 0·5 g portions of liver (Mattocks and White, 1970). After intraperitoneal dosing, where the alkaloid is absorbed rapidly, there is a fast build-up of pyrrolic metabolites in the liver. Some excess metabolite is lost within the first hour, after which the pyrrole level falls only very slowly during 24 hours and is still detectable after 2 days. A similar picture emerges after oral dosing except that absorption of the alkaloid is slower, and hence the initial surge of metabolism is absent.

In another experiment, rats were given various doses of retrorsine intraperitoneally, killed after 2 hours, and liver pyrroles measured (Fig. 13). There is a roughly linear relationship between the dose of alkaloid and the liver pyrrole

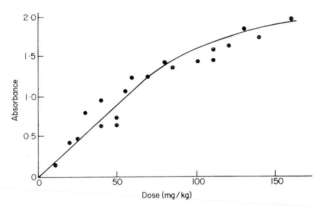

FIG. 13. Relative liver pyrrole metabolite levels in male rats two hours after various doses of the alkaloid retrorsine, given intraperitoneally

level up to well above the LD_{50} of retrorsine (40 mg/kg). Under the conditions used, no colour is produced in liver samples from untreated rats.

Table I shows the results of a series of experiments such as this, using a number of toxic alkaloids. It can be seen that the largest amounts of liver pyrroles are produced from the most toxic alkaloids. While the LD_{50}s of the alkaloids extend from about 40 to 1000 mg/kg, the pyrroles found in the liver 2 hours after an LD_{50} dose vary by a very much smaller factor. Thus there is a reasonable correlation between the acute toxicities of these alkaloids and the amounts of pyrrolic metabolites found in the liver. It should be noted that there are considerable differences in the toxicities of some of the alkaloids between male and female rats, and these are also reflected in the amounts of pyrrole metabolites.

Rats can be pretreated in ways which make the microsomal metabolizing system either more or less active. If this is followed by treatment with pyrrolizidine alkaloids, the toxicity of the alkaloids can be altered, and the

TABLE I

Acute toxicities (up to 4 days) of various pyrrolizidine alkaloids given to rats by intraperitoneal injection, compared with pyrrole metabolite levels measured in the liver 2 hours after dosing. (Arbitrary units representing absorbance of Ehrlich colour from 0·5 g of tissue)

Alkaloid	Sex	Approx. acute LD_{50} (mg/kg)	Liver pyrroles per dose of 100 mg/kg	LD_{50}
Retrorsine	M	34	1·88	0·64
	F	150	0·54	0·81
Monocrotaline	M	109	0·64	0·70
	F	230	0·42	0·96
Heliotrine	M	280	0·27	0·76
Lasiocarpine	M	77	0·87	0·67
Senecionine	M	50	1·85	0·92
Indicine	M	>1000	0·09	0·9

liver pyrrole metabolites are also different. Some examples are given in Table II. Phenobarbitone pretreatment stimulates the microsomal enzymes, and in the female rat causes a considerable increase in both the toxicity of retrorsine and in the liver pyrrole production. However, similar treatment in male rats causes a reduction in pyrrole production and in toxicity. We cannot explain this at present, but it may be that the metabolic activity is so high as to cause more complete destruction of the alkaloid. Rats fed protein deficient diets have reduced microsomal metabolic activity. Thus, in male rats fed only on sucrose for 4 days, retrorsine is reduced in toxicity and the liver pyrrole metabolites are reduced. Pretreatment with the drug SKF525A, a microsomal enzyme inhibitor, brings about a similar effect.

TABLE II

Acute (4 day) toxicities and relative liver pyrrole levels 2 hours after retrorsine (intraperitoneal) in rats given various pretreatments

Pretreatment	Sex	Approx. acute LD_{50} (mg/kg)	Liver pyrroles per dose of 100 mg/kg	LD_{50}
None	M	34	1·88	0·64
Phenobarbitone, 1 week[a]	M	65	1·67	1·10
Sucrose diet, 4 days	M	120	1·32	1·58
SKF525A[b]	M	55	1·21	0·67
None	F	150	0·54	0·81
Phenobarbitone, 1 week[a]	F	90	1·11	1·00

[a] 0·1% in drinking water.
[b] 70 mg/kg intraperitoneally, 1 hour before retrorsine.

It has been mentioned that very young rats are more susceptible than adults to the chronic toxic effects of pyrrolizidine alkaloids (Jago, 1970; Schoental, 1970). We have found higher levels of pyrrolic metabolites in the livers of young animals dosed with retrorsine, than in adults. This is shown in Fig. 14. The level in young rats is about twice that in the adult male, and it falls to the adult male level between the ages of about 17 and 27 days, after which there is a further drop in the females. The larger amounts of pyrroles produced cannot entirely account for the high sensitivity of young animals since in the latter only about one-third the dose is needed to produce similar toxic effects compared with the adult (Jago, 1970). It is likely that the greater sensitivity is associated with the presence of rapidly developing liver tissue.

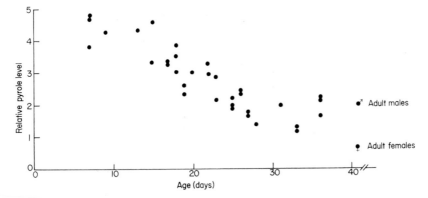

FIG. 14. Relative liver pyrrole metabolite levels in young rats of various ages, two hours after similar intraperitoneal doses of the alkaloid retrorsine

VI. TOXICITIES OF SEMISYNTHETIC AND SYNTHETIC ALKALOIDS AND PYRROLE DERIVATIVES

A. SEMISYNTHETIC AND SYNTHETIC PYRROLE DERIVATIVES

When it became clear that highly reactive pyrrolic metabolites were being formed from the alkaloids, attempts were made to prepare similar compounds and test these for toxicity (Mattocks, 1968a). The natural alkaloids can be dehydrogenated chemically by various methods to their pyrrole derivatives (dihydropyrrolizine esters) (Mattocks, 1969a; Culvenor *et al.*, 1969b, 1970a,b). These compounds, such as monocrotaline pyrrole (dehydromonocrotaline) (13) have to be administered to animals as solutions in inert solvents, such as *NN*-dimethylformamide, since they react instantly with water. For the same reason they are ineffective when given by stomach tube, since they are decomposed by the aqueous acid in the stomach. If given by intraperitoneal injection, they cause local effects such as inflammation and peritonitis, but the pyrrolic material which can be demonstrated, *post mortem* near the injection site, does not survive long enough to reach vital organs. However, when administered to rats by intravenous injection into a tail vein, the compounds are carried within a few seconds via the heart to the lungs, where they appear to be very effectively taken up (Plestina, 1972). Pyrrole derivatives given in this way cause long term lung damage, similar to that caused by alkaloids such as monocrotaline and fulvine, but at much lower doses (5–10 mg/kg compared with 50–100 mg/kg for the parent alkaloids) (Butler *et al.*, 1970). An ultramicroscopic study has been made of the resultant lung damage (Butler, 1971). To produce liver damage, the pyrrolic compound must be introduced into the bloodstream as near as possible to the liver, and a mesenteric vein, exposed by laparotomy, was chosen (Butler *et al.*, 1970). Most rats survived this treatment and appeared well after doses of 15 mg/kg of monocrotaline or retrorsine pyrroles, but autopsies revealed severe liver damage which included some of the features of pyrrolizidine alkaloid poisoning, including giant cells, but also other features such as vessel damage which could be due to the poison coming to the liver via the vessels instead of from within the liver parenchymal cells.

Two other compounds have been prepared and tested for toxicity in rats by the tail vein route. These are diacetylretronecine pyrrole (14) (Mattocks, 1970) and 1-methyl-2,3-diacetoxymethylpyrrole (15) (Mattocks, 1971b). Both these

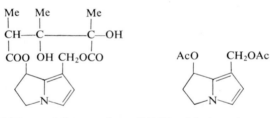

(13) Monocrotaline pyrrole   (14) Diacetylretronecine pyrrole

(15) 1-Methyl-2,3-diacetoxymethylpyrrole

compounds had the same toxic effects to the lungs as monocrotaline pyrrole (13) and at similar dose levels. The first incorporates the basic moiety from the natural alkaloids but has a simple acid moiety (acetic acid). The second compound is entirely synthetic, and has only the pyrrole ring, not the complete dihydropyrrolizine system. Thus the variations in the toxicities of the natural alkaloids are probably due more to the varying degrees to which they are metabolized to pyrrolic derivatives than to differences between the resulting metabolites (Mattocks, 1970). Hepatic megalocytosis, a characteristic chronic effect of pyrrolizidine alkaloids, has been obtained by Culvenor *et al.* (1971)

using a non-esterified pyrrole derivative. Dehydroheliotridine (11), which can be derived metabolically from some of the natural alkaloids, produced giant cells 8 weeks after being given in a single, though large (92 mg/kg) intra-peritoneal dose to 2-week-old rats. Thus even the weaker alkylating reactivity of the hydroxypyrrole is sufficient to cause long term effects in susceptible animals. On the other hand, in our experiments large intravenous doses of the related dehydroretronecine into rats failed to produce any toxic lung or liver effects. Various cytotoxic and mutagenic effects of pyrrole derivatives of the alkaloids have been investigated in cell cultures, bacteria, experimental tumours and viruses (Culvenor *et al.*, 1969a).

B. SYNTHETIC COMPOUNDS RELATED TO THE ALKALOIDS THEMSELVES

In early experiments (Schoental and Mattocks, 1960; Mattocks, 1969b) the amino-alcohol retronecine (2), the basic moiety of such toxic alkaloids as retrorsine (6) and monocrotaline (7), could be esterified with a series of simple aliphatic acids. Such diesters had the toxic properties of the natural alkaloids if the acid moieties had branched chains and 5 or more carbon atoms. Thus di-*iso*valeryl retronecine (16) was hepatotoxic to rats at doses of 200–500 mg/kg but di-*n*-butyryl retronecine (17) was not toxic at up to 1300 mg/kg. Later it was found that diacetyl retronecine (18) was hepatotoxic in high doses if adminis-tered together with an esterase inhibitor (Mattocks, 1970). It was thought that enzymic hydrolysis might be competing with pyrrole formation in the rat, so the *N*-ethyl carbamate diester of retronecine (19) was tested, since it was likely to be more resistant to esterase activity. This compound was indeed more toxic, being similar in activity to monocrotaline (Mattocks, 1971b). The next step was to prepare a similar carbamate diester (20) of an entirely synthetic amino-alcohol, 1-methyl-2,3-*bis*-hydroxymethyl-3-pyrroline. This proved even more hepatotoxic than (19) and caused toxic effects in rats similar to those of mono-crotaline at dose levels of 60–80 mg/kg. The discovery of a synthetic compound having the toxicity of a pyrrolizidine alkaloid but with only a pyrroline, not a pyrrolizidine ring, further supports the belief that pyrrole-like metabolites are responsible for toxic effects. Further, it opens possibilities for the continued study of mechanisms of toxicity by the use of other chemically prepared analogues and radioactive labelled compounds.

VII. THE NATURE OF THE TOXIC REACTION

It appears likely that toxic effects in the livers and other organs of animals poisoned by pyrrolizidine alkaloids are due to the chemical interactions of re-active metabolites with vital cell constituents. For instance, it is known that the dihydropyrrolizine esters can alkylate amines (Mattocks 1969; Culvenor *et al.*, 1970a). We have found that dehydromonocrotaline (13) can cause cross-linkage of nucleic acids *in vitro*, and dehydroheliotridine (11) has been found to

198 A. R. MATTOCKS

interact with DNA *in vitro* (Black and Jago, 1970). The similarity between the
pyrrole derivatives of the alkaloids and the antibiotic mitomycin C (21) has

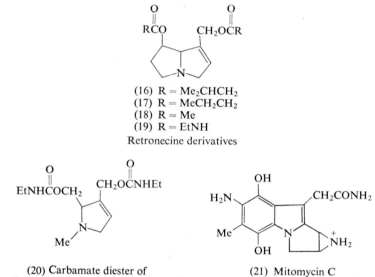

(16) R = Me₂CHCH₂
(17) R = MeCH₂CH₂
(18) R = Me
(19) R = EtNH
Retronecine derivatives

(20) Carbamate diester of
1-methyl-2,3-*bis*-hydroxymethyl-3-pyrroline

(21) Mitomycin C

been pointed out (Mattocks, 1969a; Culvenor *et al.*, 1969a). Mitomycin can
cause cross linkage of the strands in DNA through bifunctional alkylation
(Iyer and Szybalski, 1964). The toxic pyrrole derivatives are also capable of
behaving as bifunctional alkylating agents. But even if it can be demonstrated
that a compound can react with certain macromolecules in a cell, whether
nucleic acids, proteins or carbohydrates, it is another matter to discover
whether this has any connection with toxic effects. Some of the problems
connected with this kind of investigation have been discussed by Miller and
Miller (1966) in relation to various chemical carcinogens. Work in this direction with pyrrolizidine alkaloids has a very long way to go. However a start has
been made, and the tools are now becoming available.

VIII. Conclusions

We have come a long way since it was first demonstrated that animals could
die because they had grazed upon a field of *Senecio* plants many months earlier.
Studies of these plants have led to some very interesting chemistry and the
discovery of a new class of reactive compounds. The alkaloids and their
pyrrolic derivatives can be used as tools to study the biochemistry and
pathology of some intriguing liver and lung diseases. In their role as antimitotic agents they provide a basis for research on antitumour agents. Very

little is yet known about the biochemical mechanisms whereby the toxic metabolites exert their effect in the cell.

The *Senecio* story is far from being completed. It is just beginning to be interesting.

REFERENCES

Aplin, R. T., Benn, M. H. and Rothschild, M. (1968). *Nature, Lond.* **219**, 747–748.
Armstrong, S. J. and Zuckerman, A. J. (1970). *Nature, Lond.* **228**, 569–570.
Barnes, J. M., Magee, P. N. and Schoental, R. (1964). *J. Path. Bact.* **88**, 521–531.
Bingley, J. B. (1968). *Analyt. Chem.* **40**, 1166–1167.
Black, D. N. and Jago, M. V. (1970). *Biochem. J.* **118**, 347–353.
Bras, G., Berry, D. M. and Gyorgyi, P. (1957). *Lancet* **1**, 960–962.
Bull, L. B. (1955). *Aust. Vet. Jl* **31**, 33–40.
Bull, L. B., Culvenor, C. C. J. and Dick, A. T. (1968). "The Pyrrolizidine Alkaloids." North Holland Publishing Co., Amsterdam.
Butler, W. H. (1971). *J. Path.* **102**, 15–19.
Butler, W. H., Mattocks, A. R. and Barnes, J. M. (1970). *J. Path.* **100**, 169–175.
Culvenor, C. C. J., Dann, A. T. and Dick, A. T. (1962). *Nature, Lond.* **195**, 570–573.
Culvenor, C. C. J. and Smith, L. W. (1969). *Tetrahedron Letters* **41**, 3603–3606.
Culvenor, C. C. J., Downing, D. T., Edgar, J. A. and Jago, M. V. (1969a). *Ann. N.Y. Acad. Sci.* **163**, 837–847.
Culvenor, C. C. J., Edgar, J. A., Smith, L. W. and Tweeddale, H. J. (1969b). *Tetrahedron Letters* **41**, 3599–3602.
Culvenor, C. C. J., Edgar, J. A., Smith, L. W. and Tweeddale, H. J. (1970a). *Aust. J. Chem.* **23**, 1853–1857.
Culvenor, C. C. J., Edgar, J. A., Smith, L. W. and Tweeddale, H. J. (1970b). *Aust. J. Chem.* **23**, 1869–1879.
Culvenor, C. C. J., Edgar, J. A., Smith, L. W., Jago, M. V. and Peterson, J. E. (1971). *Nature New Biol.* **229**, 255–256.
Davidson, J. (1935). *J. Path. Bact.* **40**, 285–295.
Dick, A. T., Dann, A. T., Bull, L. B. and Culvenor, C. C. J. (1963). *Nature, Lond.* **197**, 207–208.
Downing, D. T. and Peterson, J. E. (1968). *Aust. J. exp. Biol. Med. Sci.* **46**, 493–502.
Iyer, V. N. and Szybalski, W. (1964). *Science, N.Y.* **145**, 55.
Jago, M. V. (1969). *Am. J. Path.* **56**, 405–421.
Jago, M. V., Lanigan, G. W., Bingley, J. B., Piercy, D. W. T., Whittem, J. H. and Titchen, D. A. (1969). *J. Path.* **98**, 115–128.
Jago, M. V. (1970). *Aust. J. exp. Biol. Med. Sci.* **48**, 93–103.
Jago, M. V., Edgar, J. A., Smith, L. W. and Culvenor, C. C. J. (1970). *Molec. Pharmac.* **6**, 402–406.
Kay, J. M. and Heath, D. (1969). "*Crotalaria spectabilis*, the Pulmonary Hypertension Plant." Charles C. Thomas, Springfield, Illinois.
Koekemoer, M. J. and Warren, F. L. (1951). *J. chem. Soc.* 66–68.
Lanigan, G. W. and Smith, L. W. (1970). *Aust. J. Agric. Res.* **21**, 493–500.
Mattocks, A. R. (1967). *Analyt. Chem.* **39**, 443–447.
Mattocks, A. R. (1968a). *Nature, Lond.* **217**, 723–728.
Mattocks, A. R. (1968b). *Analyt. Chem.* **40**, 1749–1750.
Mattocks, A. R. (1968c). *Nature, Lond.* **219**, 480.
Mattocks, A. R. (1969a). *J. chem. Soc. (C)* 1155–1162.
Mattocks, A. R. (1969b). *J. chem. Soc. (C)* 2698–2700.
Mattocks, A. R. (1970). *Nature, Lond.* **228**, 174–175.

Mattocks, A. R. and White, I. N. H. (1970). *Analyt. Biochem.* **38**, 529–535.
Mattocks, A. R. (1971a). *Trop. Sci.* **13**, 65–70.
Mattocks, A. R. (1971b). *Nature, Lond.* **232**, 476.
Mattocks, A. R. and White, I. N. H. (1971a). *New Biol.* **231**, 114–115.
Mattocks, A. R. and White, I. N. H. (1971b). *Chem-Biol. Interactions* **3**, 383–396.
McLean, E. K. (1970). *Pharmacol. Rev.* **22**, 429–483.
Miller, E. C. and Miller, J. A. (1966). *Pharmacol. Rev.* **18**, 805–838.
Plestina, R. and Stoner, H. B. (1972). *J. Path.* (In press).
Schoental, R. (1957). *Nature, Lond.* **179**, 361–363.
Schoental, R. (1959). *J. Path. Bact.* **77**, 485–495.
Schoental, R. (1968a). *Israel. J. Med. Sci.* **4**, 1133–1145.
Schoental, R. (1968b). *Cancer Res.* **28**, 2237–2246.
Schoental, R. (1970). *Nature, Lond.* **227**, 401–404.
Schoental, R. and Magee, P. N. (1957). *J. Path. Bact.* **74**, 305–319.
Schoental, R. and Magee, P. N. (1959). *J. Path. Bact.* **78**, 471–482.
Schoental, R. and Mattocks, A. R. (1960). *Nature, Lond.* **185**, 842–843.
Schoental, R., Head, M. A. and Peacock, P. R. (1954). *Br. J. Cancer* **8**, 458–465.
Warren, F. L. (1955). *In* "Progress in the Chemistry of Organic Natural Products" (L. Zechmeister, ed.), pp. 198–269, Vol. 12. Springer-Verlag, Vienna.
Warren, F. L. (1966). *In* "Progress in the Chemistry of Organic Natural Products", pp. 330–406, Vol. 24. Springer-Verlag, Vienna.
Warren, F. L. (1970). *In* "The Alkaloids" (R. H. F. Manske, ed.), pp. 245–331, Vol. 12. Academic Press, New York and London.

CHAPTER 12

Phytotoxins: An Ecological Phase of Phytochemistry

CORNELIUS H. MULLER AND CHANG-HUNG CHOU[1]

University of California, Santa Barbara, California, U.S.A.

I. INTRODUCTION

Many of the biologically active compounds produced by higher plants may find their ways into other organisms where they take greater or lesser part in a multiplicity of important physiological processes. They often attract or repel and nourish or poison animals. They may stimulate or suppress the growth of microorganisms. Some may interefere in the regulatory functions of other higher plants, sometimes stimulating growth, sometimes stopping growth altogether. When such a plant product is shown to have the potential of suppressing higher plant growth, it is called a phytotoxin, regardless of whether or not it is physiologically active relative to any animal or nonvascular plant. We do not defend this terminology on any philosophic grounds but follow the usage for purely pragmatic reasons. No functional classification of phytotoxins is possible that will agree more than generally with a chemical classification. Interference with a single enzymatic step is sufficient to control a complex process, and thus the kinds of chemical compounds involved in biotic interactions are far more numerous than the broad physiological mechanisms they affect.

We shall limit our subject to the functions of phytotoxins in controlling ecological processes characteristic of vegetation. The negative effect of one plant upon another by means of chemical products loosed into the environment

---

[1] Present address: Department of Botany, University of Toronto, Canada.

8*

is termed *allelopathy* (Molisch, 1937). We have elsewhere (Muller 1969, 1970a,b) emphasized the position of allelopathy (and the biochemical parameter upon which it depends) as one of several basic ecological processes whose chemical cause is but another major factor in the environmental complex. The chemical variable takes no precedence over light, temperature, moisture, and inorganic mineral nutrients, the traditionally recognized environmental components, It shares with each of them a part in determining the plant environment. However, it also shares in the potential for becoming a limiting factor and thus exercising control.

The first step in testing an hypothesis of allelopathic control involves investigation of all non-biochemical parameters of the environment. If the reaction upon these factors by a dominant plant is demonstrated to leave each one adequate for the needs of another plant which nonetheless is suppressed, these factors may not be invoked in an explanation of the observed suppression. Thus, certain herbs may fail to grow beneath a dense tree canopy because of shading. However, if the tree has a tall bole so that full sunlight reaches the base of the trunk, herb exclusion on the sunny side may not be ascribed to shading. When each physical factor has in like manner been proved favorable, a biochemical explanation is sought. This second step requires care that no reliance be placed upon unnatural extraction methods. Waring blendors, stills, and reflux columns have their uses, but these do not include demonstration of naturally occurring toxic phenomena. Decomposing leaf litter, rain drip or fog drip (del Moral and Muller, 1969, 1970), or the soil beneath a plant (DeBell, 1969) may yield strongly positive bioassay results without the employment of any unnatural procedures.

II. VOLATILE TOXIC TERPENES

The startlingly effective toxicity of the very common terpenes loosed into the environment by the shrub, *Salvia leucophylla* of the family Labiatae, in California has been extensively treated (Muller, 1965, 1966, 1967; Muller and del Moral, 1966). A few points need only brief mention for purposes of comparison with other toxic systems.

The observed *Salvia* phenomenon consists of the development of a zonation of herbs about *Salvia* thickets where these invade annual grassland. The interior of the shrub zone is usually devoid of herbs or even shrub seedlings. A zone 1 to 2 m in width, largely lacking plants of any sort, develops immediately adjacent to the shrub zone. Beyond this "bare" zone there characteristically occurs an "inhibition" zone in which certain grassland herbs are totally lacking and the remainder exhibit decreasingly stunted growth with greater distance from *Salvia*. Such inhibition has been recognizable 8 to 10 m distant from the shrubs during periods of normal rainfall in our droughty climate, but it may be partially obliterated by vigorous growth of grasses during several years of extremely favorable rainfall.

The pattern of development issuing from this shrub-herb interaction is strongly dependent upon the Mediterranean climate and the "annual" quality of the California grassland. The great bulk of this vegetation consists of annual species which undergo seed germination, seedling establishment, and seed production each year. Thus, the entire vegetation is subject to the hazards of seed germination and early growth of seedlings annually. Both of these functions are strongly liable to toxic influence and thus the highly visible phenomenon develops quickly and consistently.

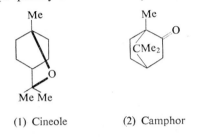

(1) Cineole   (2) Camphor

The presence of volatilized cineole (1) and camphor (2) in atmosphere adjacent to foliar branches of *Salvia leucophylla* has been detected repeatedly by gas chromatography. The same two compounds, furthermore, have been recovered from dry soil adjacent to *Salvia* shrubs. It is significant that, of the entire list of abundant terpenes recovered from the effluvium of crushed *Salvia* leaves, camphor and cineole proved the most toxic to seed germination and seedling growth in a long list of herb species. With the demonstration of rapid adsorption of terpenes upon dry soil colloids and similarly rapid solution in paraffin (comparable to the waxes of cutin in seedlings), the ecological basis was laid for the inhibition of grassland herbs adjoining *Salvia* thickets. Maximal terpene release occurs in late spring when rains have virtually ceased and herb growth has largely been completed. The terpenes remain in the dry soil over the long drought period of summer until the resumption of rain and plant growth in winter. At this time they enter seedling roots by way of the waxy cutin and suppress growth sufficiently to prevent survival under normal periodic drought stress.

Four to six weeks after the beginning of the growing season it is no longer possible to detect *Salvia* terpenes in the soil. These appear to have been degraded by soil microflora whose cycle of activity, like that of higher plants, commences with the winter rains.

The strong variation in effectiveness observed in *Salvia* toxicity to grassland herbs (Muller, 1970a) has as its basis fluctuation in soil moisture supply. Drought stress is heaviest upon those seedlings with shallow root penetration resulting from growth inhibition by toxins. Relief from drought stress locally by shading or fog drip results in increased survival of stunted seedlings. Extreme increases in soil moisture, either locally or generally as a result of increased precipitation frequency, favors growth of soil microflora and results in rapid biodegradation of soil terpenes. Thus, a single physical factor has dual effects upon the effectiveness and the persistence of a class of phytotoxic compounds. In several respects, then, volatile terpene toxicity as exhibited by *Salvia leucophylla* is unique among allelopathically significant toxin systems, although it is probably representative of a numerous class.

III. WATER-BORNE TOXINS

An ecological class of phytotoxins totally lacking in chemical coherence is recognizable by its dependence upon water as a mode of movement. Although a degree of volatility may be manifested by some members of this group, there exists no known mechanism of concentration that renders such gaseous forms ecologically significant. Solubility in water, at least in some molecular states of these compounds, is a universal condition in this group. The plants that produce them, the vegetations they form, and the physical factors under which they operate are extremely variable and result in a multiplicity of allelopathic mechanisms. A few examples will serve to illustrate the inherent differences. Although the water-soluble toxins comprise a wide variety of chemical groups, only the phenolic compounds have been widely studied in this context.

Ecological studies of complex mixtures of chemical compounds involve massive quantities of work. It is mandatory from a practical standpoint that isolation and identification procedures be rapid and highly productive. Paper chromatography fits this description and has been the principal chemical tool for studying water-soluble phytotoxins.

A. CALIFORNIA CHAPARRAL

Under the Mediterranean climate type of Southern California, the shallow soils and fissured substrates of mountain slopes support a characteristic shrub vegetation known as "chaparral". A striking characteristic of the vegetation is the complete lack of herbs beneath the shrubs, or even between shrubs, no matter how thin the canopy. This growth is subject to recurrent fires which destroy all above-ground portions of the dominant shrubs at intervals averaging about 25 years. However, in the first few growing seasons following a fire, the same area produces a luxuriant growth of many annual herb species and a multitude of shrub seedlings as well. The dominant chaparral shrubs gradually regenerate both by growth of their seedlings and by vigorous sprouting from surviving subterranean root stocks. Shrub regeneration is accompanied by gradual decrease in herb seed germination so that in 5 to 7 years the shrubs have regained dominance and herb growth has ceased. Some herb species appear only the first year and their seeds then lie dormant until the next fire, even if that is delayed 40 to 50 years. This is the widely known fire cycle of California chaparral. It has recently been suggested (Muller *et al.*, 1968) that the cessation of herb growth results from phytotoxins of the dominant shrubs which impose seed dormancy which in turn is interrupted by fire.

1. Adenostoma fasciculatum

An almost ubiquitous dominant of chaparral is *Adenostoma fasciculatum*, a member of the Rosaceae. The mode of dominance of this species was studied in detail (McPherson and Muller, 1969) with the primary purpose of assessing the role of phytotoxins in the herb response to fire. Controlled field and laboratory experiments conclusively excluded from limiting status all physical and biotic factors other than phytotoxins. The behavior of the toxic compounds is of interest here.

Since the natural field situation involves extremely limited numbers of extraction mechanisms, field and laboratory experiments were designed so as to introduce no unnatural events. Possible sources of toxins reaching the soil are rain drip and fog drip, decomposing leaf litter, and root exudation. Roots of *Adenostoma* proved, upon excavation, to be nearly vertical, leaving the surface and shallow horizons virtually unoccupied. Bioassays of fresh aqueous root extracts failed to show inhibition of herb seedling growth. Leaf litter proved equally benign. Fog drip and rain drip, on the other hand, whether collected under natural conditions or artificially produced in the field or laboratory, contained effective toxins. These compounds appear to be deposited upon the leaf surface in the course of normal metabolic function. They accumulate

TABLE I

Compounds isolated from aqueous extracts of *Adenostoma* foliar branches and alkaline ethanol extracts of *Adenostoma* soil

Compounds[b]	Foliage	Soil
Arbutin (3)[a]	+	
Hydroquinone[a]	+	
Phloridzin	+	
Umbelliferone[a]	+	
Ferulic acid (4)[a]	+	+
p-Coumaric acid[a]	+	+
Syringic acid (6)[a]	+	+
Vanillic acid[a]	+	+
p-Hydroxybenzoic acid (5)[a]	+	+

[a] Yielded outstandingly toxic chromatogram segments.
[b] Arranged generally in the order of decreasing abundance in foliage and increasing abundance in soil.

during periods of atmospheric drought and are rapidly depleted by the leaching action of as little as 5 or 10 mm of rain. The deposition of the toxic materials in the soil results in extreme reduction of rate of seed germination and in seedling growth, thus laying the basis for full dominance by the adult shrubs.

The leachate of *Adenostoma* foliage has yielded 9 identifiable phenolic compounds (Table I), comprising glycosides and phenolic acids (e.g. 3–6). All of

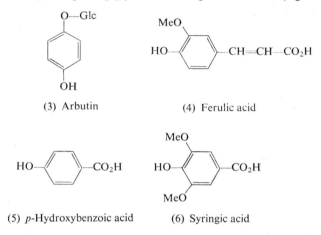

(3) Arbutin

(4) Ferulic acid

(5) *p*-Hydroxybenzoic acid

(6) Syringic acid

these showed toxicity to seedling growth and most of them inhibited germination in bioassays involving "seeds" of *Lactuca sativa* (McPherson *et al.*, 1971). In addition, removal of all phenolic compounds by base extraction left a residue of several glycosides with toxic moieties which have not yet been identified.

Extraction from soil beneath *Adenostoma* shrubs was undertaken by means of a modification of the alkaline ethanol method of Wang *et al.* (1967). This yielded 5 of the 9 phenolic compounds found in aqueous foliar leachate. However, these compounds were by no means evenly distributed. They were most abundant beneath shrubs with a thin layer of leaf litter and least abundant in openings between shrubs lacking any litter at all. Nevertheless, both kinds of soil surfaces are characteristically devoid of herb growth. It therefore seems doubtful that the extractable phenolic fraction of the soil organic complex is the sole or even the principal agent of toxicity.

One of the most dynamic qualities of *Adenostoma* toxicity is its short duration in the environment. The removal of above-ground parts of shrubs without soil disturbance permits the same flush of herb growth in the next growing season that one observes following fire. The clearing shown in Fig. 1 was photographed in its first year following shrub removal. A lag period of only 10 to 12 weeks characterizes the herb response to mechanical shrub removal. Clearly, continuing suppression of herbs within the shrub stands is dependent upon periodic renewal of soil toxins, such as occurs with each instance of foliar leaching by rain or fog drip.

FIG. 1. Herb and shrub seedling response to removal of *Adenostoma* foliar crowns photographed at the end of the first growing season following clearing

2. *Arctostaphylos glandulosa* var. *zacensis*

In a study now nearing completion (Chou and Muller, 1972) we examined the Ericaceous shrub *Arctostaphylos glandulosa* var. *zacensis* as it relates to the fire cycle of California chaparral. In spite of the similarity of its herb control to that by *Adenostoma*, there have emerged some differences that indicate strong deviation between the two shrubs in the persistence and fate of their biochemical products.

Arctostaphylos forms dense pure stands in some localities of the San Rafael Mountains. The lack of herbs in these stands might at first appear to be the consequence of the dense shade cast by the broad leaves and interlocking crowns of the shrubs. However, the occurrence of small openings, both natural and man-made, permits the several physical factors to be fully evaluated relative to failure of herb growth. Openings several meters in breadth, so situated that surface drainage from *Arctostaphylos* thickets flows over them, are totally lacking in herb growth (Fig. 2). Similar openings, where drainage from shrub areas fails to reach them, are occupied by several species of annual herbs. Both kinds of clearings receive undiminished insolation and the same adequate precipitation. Mineral nutrient analyses revealed no significant differences between soils of the several situations in the general area.

The toxicity of *Arctostaphylos* tissues was established by bioassay of a series of aqueous extracts. Although living roots, living foliage, fallen fruit, exfoliated bark, and leaf litter all yielded toxic aqueous extracts, the greatest quantities of the most effective toxins were encountered in freshly fallen leaves. Rain drip from living foliage showed toxicity similar to that of *Adenostoma*, but both the

Fig. 2. An old opening in *Arctostaphylos* resulting from bulldozer scraping. Cessation of seed germination is correlated with surface drainage from the upper shrub thicket

root and litter toxicity of *Arctostaphylos* contrasts strongly with the lack of toxicity in *Adenostoma* roots and litter. We have devoted most of our attention to toxins of foliar origin because this source looms large in the total toxin supply.

Fig. 3. Lack of herbs in an artificial clearing within an *Arctostaphylos* thicket photographed at the end of the second growing season following shrub removal

Bioassays revealed extreme toxicity in freshly fallen leaf litter as well as significant toxicity of soil beneath *Arctostaphylos* shrubs. A shrub clearing experiment (Fig. 3) identical in procedure to that of *Adenostoma* yielded evidence of a basic difference between the two shrub species. In new *Arctostaphylos* clearings herb production at the end of the first growing season stood just above zero while the corresponding value for the *Adenostoma* clearing was over 1000/m² (Fig. 4). Even at the end of the second growing season the *Arctostaphylos* clearing had produced only about 36 seedlings per m². These differences suggest a long persistence of *Arctostaphylos* phytotoxins in contrast to the rapid losses in *Adenostoma* areas cleared of the shrub foliar sources of toxins.

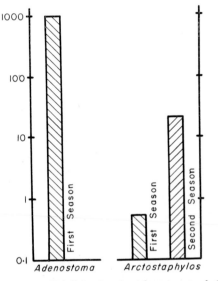

FIG. 4. Seedling counts in new artificial clearings in *Adenostoma* and *Arctostaphylos* thickets (illustrated in Figs. 1 and 3)

Aqueous leaf leachates of *Arctostaphylos* yielded 8 identified and 2 unknown phenolic compounds (Table II). Alkaline ethanol extracts of *Arctostaphylos* soil contained 4 of these as well as an additional 3 phenolic acids not found in leaf extracts. A comparison of the leaf leachates of the two shrub species reveals great dissimilarity between the phenolic compounds. However, when the soil extracts are compared, the lists of recognized phenolic compounds are much more similar. This casts some doubt upon the sufficiency of extractable soil phenolic acids to explain toxicity in two soils so functionally different.

One of the most significant characteristics of the problem lies in the relation of humus content of soil to extractability of its phenolic acid content. We have repeatedly had much greater success in isolating these acids from sites with minimal humus content than from the more normal areas of heavy leaf litter.

TABLE II

Compounds isolated from aqueous extracts of *Arctostaphylos*
leaves and alkaline ethanol extracts of *Arctostaphylos* soil

Compounds[b]	Leaves	Soil
Arbutin (3)	+	
Hydroquinone[a]	+	
Gallic acid	+	
Unknown A	+	
Chlorogenic acid	+	
Protocatechuic acid	+	
Ferulic acid (4)	+	+
Unknown B	+	+
p-Hydroxybenzoic acid (5)[a]	+	+
Vanillic acid[a]	+	+
Syringic acid (6)[a]		+
o-Coumaric acid		+
p-Coumaric acid		+

[a] Yielded outstandingly toxic chromatogram segments.
[b] Arranged generally in the order of decreasing abundance in leaves
and increasing abundance in soil.

An experiment was performed in which milled fresh leaf litter was added to soils characterized by contrasting humus content. These soils were kept moist in the greenhouse and periodically sampled for extractability of *Arctostaphylos* compounds and for toxicity. The toxicity imparted to each soil by the milled leaf litter gradually diminished in bioassays over a 6-month period, but it disappeared most rapidly in soils with a high natural organic content (29%) and remained at a significant level in similar soils with only 5% organic matter. Some compounds, such as p-hydroxybenzoic acid (5) and syringic acid (6), while initially readily leachable by cold water, eventually appeared only in alkaline ethanol extracts. Thus, what seemed to be a disappearance of compounds was very likely no more than a complexing with humic acid and organic colloids.

B. SUBHUMID DECIDUOUS FOREST

The great complexities of the more humid vegetations of the world pose serious problems in the study of allelopathy. Even its recognition is rendered difficult and, except in unusually favorable circumstances, its conclusive demonstration may be impossible. The great concentration of attention upon mesic, temperate zone landscapes has therefore resulted in a large population of professional plant ecologists who have not been convinced that allelopathy could be operative in their environments, regardless of how strikingly developed it is elsewhere. Loss of toxins by rapid soil leaching and biodegradation under conditions of abundant rainfall are frequently cited as reasons for skepticism. Therefore, more than usual importance attaches to positive evidences of allelopathy in situations of high humidity.

1. *Quercus falcata* var. *pagodaefolia*

Among several fairly conclusive instances of effective allelopathy in sub-humid situations, one of exceptional importance has recently been studied in a wet floodplain forest in South Carolina. Hook and Stubbs (1967) described the

Fig. 5. Undergrowth beneath conserved *Liquidambar styraciflua* during the third growing season following forest cutting

regeneration of this forest following cutting in which seed trees of the dominant species were left standing at a total density of 7 per acre. The dense growth of over 40 species of shrubs, vines, and tree saplings extended beneath the crowns of most trees without visible change. Thus, *Liquidambar styraciflua* (Fig. 5), extensively sampled, showed virtually no effect on the undergrowth beneath its crown in the third year after cutting of the surrounding forest. In contrast to this, however, *Quercus falcata* var. *pagodaefolia* typically inhibited the growth of woody plants so severely as to produce a grassy sward with only a few stunted tree seedlings (Fig. 6). The tall boles of the trees permitted full sunlight to reach the ground. Annual rainfall in excess of 1200 mm and periodic flooding and silt deposit provided uniformly an abundance of water and mineral nutrients. The coincidence of undergrowth inhibition with the drip line of the foliar crowns of the trees indicated that a phytotoxin leached from the leaves of *Quercus* was responsible for the inhibition.

This phenomenon was studied experimentally at the same site by DeBell (1969). He demonstrated in both field and pot experiments that soil beneath *Quercus* inhibited germination and growth of both *Quercus* and *Liquidambar*, although *Liquidambar* was considerably the more sensitive. In contrast, soil beneath *Liquidambar* proved incapable of such inhibition, growth being equal to that in soil from areas unoccupied by any tree species. He furthermore showed aqueous leachate of *Quercus* foliage to be toxic in bioassays.

FIG. 6. Suppressed undergrowth beneath conserved *Quercus falcata* var. *pagodaefolia* during the third growing season following forest cutting

Chromatography of concentrated leachate revealed two compounds, the more abundant and toxic of which he identified as salicylic acid (7).

(7) Salicylic acid

The toxic quality of soil beneath *Quercus* is quickly lost after cutting. Areas formerly occupied by *Quercus* showed no residual differences in regeneration by the third growing season. This suggests a short survival period for the toxic compounds and a dependence of the allelopathic phenomenon upon continuing phytotoxin release.

IV. ROLE OF PHYTOCHEMISTRY IN ALLELOPATHIC STUDIES

Allelopathy is fundamentally an ecological phenomenon (Muller, 1969); it must be understood in that context before it can safely be considered relative to such applied plant sciences as agronomy, silviculture, or range management. Similarly, the application of phytochemical techniques to allelopathic studies is primarily aimed at ecological analysis and the reconstruction of chemical mechanisms of ecological significance. The discovery of new and exciting compounds fails to inspire the ecologist who still needs to know how some commonplace phytotoxin acts or fails to act in the interrelation between two species.

With the establishment by means of laboratory and field experimentation that a particular instance of suppression is biochemical in nature, the ecologist frequently reaches the limit of his investigations. The multiplicity of toxic compounds produced by decomposition (Curtis and Cottam, 1950) or by several toxic species suppressing the seed germination of a single tree species (Brown, 1967) renders toxin identification an expensive undertaking. Should the ecologist persist to the point of identifying the several potentially effective toxins in each situation, he then encounters additional complications likely to be beyond his expertise.

A. IMPLICATION OF SPECIFIC PHYTOTOXINS

It is often much less difficult to establish the chemical quality of an instance of inhibition of one plant by another than to assign ecological significance to one of a complex of phytotoxins. The greater the number of toxins and the higher their potency, the greater the confidence one has in biochemical mechanisms of suppression rather than shading, competition for soil moisture, or other nonchemical interference. However, each toxin in such a complex casts some doubt upon the importance of every other one. There always exists the possibility, also, that some highly potent and extremely obscure toxin is being overlooked. A number of additional criticisms may quite legitimately be offered in such instances.

When the same series of phenolic acids is encountered repeatedly in soils beneath a wide variety of plants (grasses, broad-leafed herbs, shrubs and trees of diverse phylogenetic affiliation), these compounds lose much of their appeal as ecologically significant phytotoxins. They may indeed be the responsible toxic agents if they exhibit significantly higher concentrations in those soils least hospitable to plant growth. They may, however, be no more than the most stable products of lignin decomposition and scarcely peculiar to allelopathic plant species at all.

The possibility of a highly potent, ephemeral toxin is the most difficult to dismiss. Should such a toxin be more or less continuously liberated and quickly transformed to a relatively stable common compound, it may never be detected in its active state. Winter (1961) has stressed that such continuous production and decomposition implies equal availability for absorption by higher plants and by microbial decomposers so that such a toxin could be effective without ever building up great concentrations in the soil.

While the probability is very high that multiple toxins act synergistically, or at least additively, this lack of specificity leaves the investigator at best aesthetically dissatisfied. Such a phenomenon may lend itself to *in vitro* experimentation, but its confirmation in the field would scarcely be practical. Thus, one continues to search for more definite evidence of a specific phytotoxin, although in reality the inhibitory quality of the biochemical parameter may lie in the combined actions of a large number of individually inadequate toxic compounds.

B. ECOLOGY OF PHYTOTOXINS

We have emphasized in this discussion the ecology of plants in which biochemical effluvia play an important part. It should be pointed out, however, that there exists not only an ecology of plants but also an ecology of chemical compounds such that these may be accumulated or dissipated, stabilized or transformed, rendered more toxic or less so. To the extent that plant ecology comprises a biochemical environmental parameter, it is dependent upon the outcome of the flux that we call the ecology of biochemical effluvia.

We know that certain molecular rearrangements commonly occur following release from a plant source. For instance, arbutin is hydrolysed enzymatically in soil with the production of hydroquinone and then benzoquinone, this resulting in markedly increased toxicity. Complexing of phenolic compounds with soil colloids is strongly suggested by several lines of evidence. Thus, *Eucalyptus camaldulensis* effectively inhibits herbs on clay loam but is totally ineffective on nearby sand (del Moral and Muller, 1970). Wang *et al.* (1971) showed that phenolic acids added to soil disappeared differentially, *p*-hydroxybenzoic acid being more persistent than ferulic, syringic, *p*-coumaric, and vanillic acids, for instance. They showed that additions of small quantities of one acid results in the release of others, *p*-hydroxybenzoic, *p*-coumaric, and vanillic acids producing this release while ferulic and syringic acids did not. Addition of large quantities of phenolic acids resulted in fixation of extractable acids, apparently by incorporation into humic acid. We have found, as just described, that the entire complex of phenolic acids and quinones in milled fresh leaf litter of *Arctostaphylos* disappeared from moist soil with the passage of time, their detection in aqueous extracts of the soil aliquots being impossible after six months. More significantly, however, the rate of diminution of extractable compounds increased greatly with increasing humus content of the original soil. We conclude from this that complexing with humic materials is the basis of the apparent disappearance of these compounds. Since similar mixtures of phenolic acids from *Eucalyptus* were more active and effective in a clay loam than in sand, one is inclined to conclude that complexing with inorganic soil constituents may serve to concentrate these toxins after the manner of column chromatography and thus increase rather than decrease their effective toxicity.

It is entirely likely that ecologists have largely overlooked whole series of phytotoxins whose behaviour is even less familiar than that of terpenes and phenolic compounds. Although absinthin has been linked to allelopathic activity (Bode, 1940), the vast number of biologically active alkaloids has generally not been implicated. Similarly, the flavonoids, steroids, and the organic sulfur compounds have been suspected but otherwise neglected. Thus, the need that exists is a double one comprising both the ecology of known released toxins and the discovery of the identities of the less familiar ones whose significance may be equally as great.

Because soil chemists have generally concentrated their efforts upon humic materials, no group of chemists has yet become deeply involved in allelopathic problems. The ecology of plant effluvia is therefore a field largely without workers and constitutes a challenge far beyond the capacities of the plant ecologists who are only now discovering it. The resulting lacuna would much better be filled by those phytochemists who have both an understanding of holistic ecological principles and a full appreciation of the beautiful complexity these imply.

ACKNOWLEDGEMENTS

This work was supported by grant GB-14891 from the National Science Foundation. We are furthermore indebted to Dr J. K. McPherson for the photograph in Fig. 1, to the Ecological Society of America for permission to reproduce this (from McPherson and Muller, 1969), and to the Southeastern Forest Experiment Station for Figs. 5 and 6 from the work of Hook and Stubbs (1967).

REFERENCES

Bode, H. R. (1940). *Planta* **30**, 567.
Brown, R. T. (1967). *Ecology* **48**, 542.
Chou, C.-H. and Muller, C. H. (1972). *Am. Midl. Nat.* (in press).
Curtis, J. T. and Cottam, G. (1950). *Bull. Torrey Bot. Club* **77**, 187.
DeBell, D. S. (1969). Dissertation, School of Forestry, Duke University, Durham, North Carolina.
del Moral, R. and Muller, C. H. (1969). *Bull. Torrey Bot. Club* **96**, 467.
del Moral, R. and Muller, C. H. (1970). *Am. Midl. Nat.* **63**, 254.
Hook, D. D. and Stubbs, J. (1967). *U.S. For. Serv. Research Notes*, SE-70.
McPherson, J. K. and Muller, C. H. (1969). *Ecol. Monogr.* **39**, 177.
McPherson, J. K., Chou, C.-H. and Muller, C. H. (1971). *Phytochemistry* **10**, 2925.
Molisch, H. (1937). "Der Einfluss einer Pflanze auf die andere-Allelopathie." Fischer, Jena.
Muller, C. H. (1965). *Bull. Torrey Bot. Club* **92**, 38.

Muller, C. H. (1966). *Bull. Torrey Bot. Club* **93**, 332.

Muller, C. H. (1967). *PflKrankh.* (*PflPath.*) *PflSchutz* **74**, 333.

Muller, C. H. (1969). *Vegetatio* **18**, 348.

Muller, C. H. (1970a). *Recent Adv. Phytochemistry* **3**, 105.

Muller, C. H. (1970b). *In* "Biochemical Coevolution" (K. L. Chambers, ed.), pp. 13–31. Oregon State Univ. Press, Corvallis, Oregon, U.S.A.

Muller, C. H. and del Moral, R. (1966). *Bull. Torrey Bot. Club* **93**, 130.

Muller, C. H., Hanawalt, R. B. and McPherson, J. K. (1968). *Bull. Torrey Bot. Club* **95**, 225.

Wang, T. S. C., Yeh, K.-L., Chen, S.-Y. and Yang, T.-.K. (1971). *In* "Biochemical Interactions among Plants", pp. 113–130. National Academy of Sciences, Washington, D.C.

Wang, T. S. C., Yang, T.-K. and Chuang, T.-T. (1967). *Soil Sci.* **103**, 239.

Winter, A. G. (1961). *Symp. Soc. exp. Biol.* **15**, 229.

CHAPTER 13

Phytoalexins

B. J. DEVERALL

A.R.C. Unit on Plant Growth Substances and Systemic Fungicides,
Wye College (University of London), nr. Ashford,
Kent, England

This review is mainly concerned with parasitism on plants by fungi but there will also be some account of parasitism by bacteria. Successful parasitism is likely to be prevented by barriers both on the surface of and inside many healthy plants. There are also distinct possibilities that the special nutrient requirements of parasites may not be provided by some plants. The present account is concerned with the existence in plants of defensive systems which are activated by infection, and this leads directly to the concept of phytoalexins. The subject of phytoalexins has been reviewed by Cruickshank (1963) and, with an emphasis on the associated biochemical changes in diseased plants, by Kuć (1968). The emphasis adopted in this paper is that of the biologist interested in the extent to which phytoalexin accumulation in diseased tissues can account for some of the common phenomena observed as plants resist attempted infections.

I. THE CONCEPT OF PHYTOALEXINS

The concept was formalized by Müller and Börger (1941) following studies on resistance to potato blight disease where distinct physiological races of the causal organism *Phytophthora infestans* can be differentiated by their reactions with different potato varieties. Müller had become well aware of the importance of hypersensitivity in resistance to potato blight. Hypersensitivity, as a phenomenon, has been known since the beginning of the century when observations showed that attempted infection by parasites of the wrong hosts often led to an hypersensitive response (Ward, 1905; Stakman, 1915). This hypersensitive response was observed as cytoplasmic granulation and browning in the host cell and resulted in the limitation of the parasite to the invasion site. In addition to considering the importance of the hypersensitive process in resistance to potato blight, Müller and Börger performed experiments which led them to suggest that a process somewhat akin to that of acquired immunity in animals could be achieved in plants. In these experiments, Müller and Börger used cut surfaces of potato tubers and different races of the fungus *Ph. infestans*. Basically they were able to show that attempted infection by an incompatible race of *Ph. infestans*, which caused an hypersensitive reaction on the surface of potato tuber, rendered that tuber resistant to subsequent infection by a normally pathogenic race. They showed that this cross protection was restricted to treated parts of the tuber only and that it was effective against unrelated fungal pathogens of potato. The process of cross protection could have been caused in many ways, such as direct chemical and/or physical interference between the two races of the fungi. However, Müller and Börger believed they had demonstrated acquired resistance by the host plant because the protection was effective in the layer of tuber below the hypersensitive tissue. By cutting away the cells which had undergone the hypersensitive response to the incompatible race of the fungus, they showed that the tissue immediately below retained resistance to the compatible race. Furthermore, as there was no antagonism between the protecting race and the pathogenic race in varieties of tuber susceptible to both races there was no indication of direct chemical interaction between the two races of the fungus. They concluded that the potatoes had acquired resistance because some defence system had been activated during attempted infection by the incompatible race.

The term phytoalexin (derived from the Greek) was proposed by Müller and Börger in 1941 to mean a "warding-off" compound produced by a plant. Müller and Börger made a number of postulates about the nature of these hypothetical compounds based on biological observations and theoretical considerations before there was any evidence of chemical entities in their work. The most important postulate was that phytoalexins are substances inhibiting development of fungi in hypersensitive tissues, formed or activated only when the host cells came into contact with parasites. It was suggested that phytoalexins are produced in the process of death of the host cell, and that they are non-specific in toxicity towards fungi. It was also suggested that similar responses occurred in resistant and susceptible tissue and that a basic difference between resistant and susceptible plants was related to the speed of formation of phytoalexin. The precise meaning attached to the word "speed" is not clear from these early writings and it is of historical and academic interest in the light of recent work discussed later.

II. ORCHID TUBERS, CROSS PROTECTION AND ORCHINOL

Müller and Börger were not the first to consider the possibility of plants acquiring a resistance towards diseases. This subject has been actively discussed for many years and earlier work has been reviewed by Chester (1933). The most important earlier study was that of Bernard (1909, 1911) who was able to cross protect excised orchid embryos against pathogenic fungi by previous exposure to avirulent fungi. Many of these experiments are difficult to evaluate because of the meagre data presented, but Nobécourt (1923) confirmed Bernard's finding and also discovered that antifungal substances diffused from pieces of orchid tuber plated on agar near to pathogenic fungi. There was some dispute at the time, witness the work of Magrou (1924), whether in fact these substances were produced from healthy tubers or whether they were produced by the orchid tissue in response to substances arising from the fungus, and this conflict was not clearly resolved until Gaümann and his associates re-examined this work (Gaümann and Jaag, 1945). Subsequent studies by Gaümann and Kern (1959a,b) showed that there was a negligible content of antifungal materials in healthy orchid tubers but that high concentrations of substances subsequently identified as orchinol (1) (Hardegger et al., 1963) and the related hircinol (Gaümann, 1964) were produced in different species of tuber after infection only. Orchinol production was induced in the orchid Orchis militaris by many fungi, and also under the influence of certain toxic chemicals.

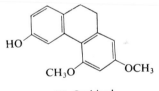

(1) Orchinol

Gaümann and his associates performed many experiments on the formation and activity of orchinol in orchid tubers in response to infection by different organisms; some of their most important findings can be summarized as follows. Gaümann and Hohl (1960) showed that, by infecting cores of excised orchid tuber at one end with the pathogenic fungus Rhizoctonia repens and allowing these to incubate for periods of time, they were able to extract orchinol from different portions in the core. For example, 8 days after infection, they obtained yields of 920 μg/g fresh tuber from positions immediately adjacent to the fungus. Much lower concentrations were present 1 cm away. Gaümann et al. (1960) studied the biological activity of orchinol towards many fungi and, although their bioassay system prevented absolute precision in estimating the activity of orchinol, they were able to show that it was inhibitory towards many mycorrhizal fungi at concentrations of the order of 20–200 μg/ml. It was not very active against the pathogenic fungus *Rh. repens*, and about 2000 μg/ml was needed to inhibit this fungus. Therefore, although much remains to be learned of the role of orchinol in the process of cross protection, there is no doubt that orchids are stimulated to make potent antifungal substances in response to infection by fungi.

The work of Gaümann and his associates was proceeding throughout the period when Müller was interested in phytoalexins, but Gaümann and Kern (1959b) chose not to use the term phytoalexin, preferring to use the term "antibody" for orchinol. This does not seem to be particularly appropriate because of the precise use of the word antibody in medicine, as a protein specifically able to counter an inducing antigen, whereas orchinol was a non-proteinaceous substance formed readily in response to many sorts of stimuli and effective against many different types of organisms.

III. Pisatin and Phaseollin as Phytoalexins

Müller never demonstrated a chemical entity associated with hypersensitivity and cross protection in potato tubers, and his demonstration that a phytoalexin might be a specific substance came in 1958 after work with the fruit pathogen *Sclerotinia fructicola* placed as spores in the seed cavities of opened bean pods, *Phaseolus vulgaris* (Müller, 1958). This fruit pathogen is not a normal pathogen of beans. A number of observations were made of what happened following inoculation of bean with *Sclerotinia*. Hypersensitive flecking of the host cells inside the seed cavity occurred within 24 hours of placing spores in the cavities. Furthermore, the affected cells took up the stain Rhodamine B increasingly strongly 12 hours after placing spores, indicating cellular damage at this time. Müller was able to show a change to a more acid pH in the infected cells 14 hours after putting down infection droplets. The important finding was that when the infection droplets were collected, combined, made spore free and tested for their effects on new spores *in vitro*, they became increasingly antifungal 14 hours after they had been placed in the seed cavities. They were completely fungistatic towards test organisms 24 hours after spores had been placed in seed cavities of pods. Müller showed that light petroleum removed an antifungal compound from the combined infection droplets leaving a water phase which was highly stimulatory to test spores. Control droplets of water from healthy seed cavities were highly stimulatory to test spores. Thus, Müller had evidence of a chemical entity extractable by light petroleum acting apparently as a single fraction and responsible for the antifungal activity generated in infection droplets while cells were undergoing hypersensitivity. At this stage, Müller retired and further work was continued by Cruickshank and colleagues in Canberra, Australia, where Müller had been working.

The first demonstration of the nature of a substance produced in a leguminous pod seed cavity in response to infection was not from bean research. For reasons of convenience, Cruickshank and his colleagues worked first with pea pods, using the same fruit pathogen *Sclerotinia fructicola*. They were able to obtain, in just the same way as Müller had done with bean, a highly active antifungal fraction, which partitioned into light petroleum (Cruickshank and Perrin, 1960). The substance responsible for the antifungal action in these droplets was isolated, crystallized and subsequently identified as pisatin (2) (Perrin and Bottomley, 1962). Soon after this it was shown that the substance produced in bean seed cavities was a closely related compound phaseollin (3) (Cruickshank and Perrin, 1963a). The structure of phaseollin was described by Perrin (1964). Thus pisatin and phaseollin were identified as components of infection droplets in pea and bean respectively.

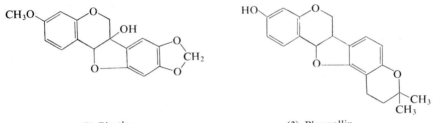

(2) Pisatin   (3) Phaseollin

Pisatin has been the subject of many research papers. Cruickshank and Perrin (1963b) compiled data on changes in pisatin concentration in droplets incubated in seed cavities of pea pods. Most fungi, whether pathogens of pea or not, but none of the bacteria tested, caused appreciable accumulations of pisatin between 12 and 72 hours after being placed in seed cavities. Interestingly, one pea pathogen, *Septoria pisi*, caused rather little pisatin to accumulate (10 μg/ml after 72 hours), but two others *Ascochyta pisi* and *Fusarium solani* var. *martii* f. *pisi* caused much more to accumulate (the order of 100 μg/ml). Accumulation in response to a particular infecting organism was much affected by age of pea pods (old pods being least productive) and by concentration of inoculum, there being an interesting but unexplained decline in pisatin accumulation at higher spore concentrations. The possibility that pisatin was the product of an endogenous microbial contaminant of pea pods, released in its activity following attempted infection, was rendered unlikely by the demonstration that spore-free germination exudates and a number of heavy metal ions caused pisatin formation. Further evidence of chemical stimulation of pisatin formation was provided by Perrin and Cruickshank (1965), Schwochau and Hadwiger (1968) and Bailey (1969). Furthermore, Bailey (1970) has shown the production of pisatin by pea callus tissue in axenic culture, possibly largely in response to components of coconut milk provided in the culture medium. Despite the abundant publications on pisatin, it is difficult to assess its significance in the resistance of pea. Cruickshank (1962) published extensive lists of the effects of pisatin on growth of many fungi, yeasts and bacteria, and from these it is possible to see that many fungi, including *Sc. fructicola*, are strongly inhibited by 100 μg/ml pisatin, a concentration of pisatin often approached after 72 hours incubation of the same fungi in seed cavities under optimal conditions (Cruickshank and Perrin, 1963b). Except for *Septoria pisi*, most of the pea pathogens tested were less sensitive to pisatin; particularly striking was the much lower sensitivity of an isolate of *Ascochyta pisi* from pea than of one from broad bean. In order to assess the importance of pisatin in resistance, it is desirable to know the concentrations of pisatin inside pea tissues at significant stages during the success or failure of particular organisms to grow inside the tissues.

Importance has been attached (Cruickshank and Perrin, 1963b; Cruickshank, 1963) to the amount of pisatin produced following infection, and the concentrations detected in infection droplets in seed cavities of pods have been taken as a direct indication of amounts "formed" in particular infections. The writer has deliberately used the term "accumulated" in place of the word "formed" when discussing these data. Accumulation would be an indication of the amounts resulting from formation by the host, and degradation by the pathogen, as well as other types of loss. There is evidence that fungi, pathogens of pea in particular, can degrade pisatin (De Wit-Elshove, 1968, 1969; Christenson, 1969; De Wit-Elshove and Fuchs, 1971). Other relevant evidence will be discussed later, partly in the light of the writer's failing to find a phytoalexin in certain infection droplets in pods of *Vicia faba* and yet obtaining substantial quantities from tissues in advance of the same type of infection (Deverall, 1967; Deverall and Vessey, 1969; Letcher *et al.*, 1970).

Phaseollin has also been the subject of numerous publications, and most of the phenomena described above with respect to pisatin have been replicated with phaseollin (Cruickshank and Perrin, 1971). Of particular interest is the isolation of a peptide, called monilicolin A, from mycelium of *Sclerotinia fructicola* (Cruickshank and Perrin, 1968). This peptide caused a significant accumulation of phaseollin in droplets in seed cavities of bean when applied at a concentration of 0·02 μg/ml ($2\cdot5 \times 10^{-9}$ M), and caused no visible damage to bean tissue at concentrations up to 10 μg/ml. As distinct from work with pisatin, several attempts have been made to assess the significance of phaseollin during the progression of different pathogens through bean tissue; this will be discussed later in Section V.

IV. A RANGE OF CHEMICAL COMPOUNDS AS PHYTOALEXINS

Throughout the long period from the time when the existence of phytoalexins was proposed to the time when pisatin was identified, others revealed the production of antifungal compounds in infected plants. In Japan, Hiura (1943) detected an antifungal compound in sweet potato roots (*Ipomoea*) infected with the fungus *Ceratocystis fimbriata*, and this compound was identified as ipomeamarone (4) by Kubota and Matsuura (1953). Ipomeamarone accumulated in substantial amounts in sweet potato tissue treated with $HgCl_2$ as well as after infection (Uritani *et al.*, 1960), and it has been suggested that ipomeamarone may have a role in disease resistance of sweet potato (Uritani and Akazawa, 1955) though critical evidence for this idea is lacking. In the U.S.A., Condon and Kuć (1960) isolated a fungitoxic compound from carrot roots infected with *C. fimbriata*, which is a pathogen of sweet potato and not of carrot. The highest yields obtained were 620 μg/g which exceeded the concentrations which were highly inhibitory to fungal growth in liquid carrot–dextrose medium. Condon and Kuć (1962) confirmed that the compound was a substituted isocoumarin (5) found previously by Sondheimer (1961) in carrot roots stored at $0°C$. This isocoumarin thus seems to be produced in biologically significant amounts in carrot roots subjected to abnormal conditions, including infection.

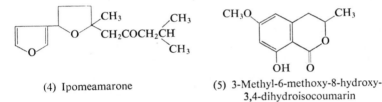

(4) Ipomeamarone

(5) 3-Methyl-6-methoxy-8-hydroxy-
3,4-dihydroisocoumarin

Of great interest has been a recent Japanese discovery of an antifungal substance produced by potato tubers during an hypersensitive response to *Phytophthora infestans*. The terpenoid, rishitin (6) (Katsui *et al.*, 1968), was found to accumulate at a concentration of 120 μg/g potato tuber two days after slices had been inoculated with an incompatible race (Tomiyama *et al.*, 1968). After comparable periods of incubation following inoculation with a compatible race, which did not cause hypersensitivity, only 0·44 μg rishitin/g tuber slice were found. A trace of rishitin was detected in uninoculated tuber slices. The ED_{50} of rishitin to *Ph. infestans* was 45 μg/ml in these experiments, thus suggesting that rishitin might be quite effective at the concentrations detected in hypersensitive tissue. Varns *et al.* (1971) confirmed the production of rishitin in hypersensitive reactions between several different races of the fungus and numerous varieties of potato tuber and described the co-incident formation of another terpenoid, phytuberin, $C_{17}H_{26}O_4$. Varns and Kuć (1971) showed not only that a compatible race failed to cause rishitin accumulation, but also that prior inoculation with such a race diminished the amount of rishitin formed when tuber slices were inoculated with an incompatible race some hours later. They suggested that compatible races had the ability to suppress both the hypersensitive response and rishitin accumulation; this idea will be discussed further at the end of this review. It seems very likely that rishitin is the entity envisaged by Müller and Börger (1941) when they proposed the existence of a phytoalexin in the potato.

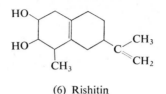

(6) Rishitin

In the last few years, there have been many reports of antifungal substances being detected in different plants after infection (see review by Kuć, 1968). Some other recent examples include the demonstration by Smith (1971) of the accumulation of numerous antifungal compounds in seed cavities of pods of many legumes, sometimes in response to attempted infection and sometimes in response to injury alone, and the identification of medicarpin and safynol as phytoalexins in alfalfa and safflower respectively. Medicarpin (7) was identified by Smith *et al.* (1971) as a substance which accumulated in infection droplets containing the corn pathogen *Helminthosporium turcicum* on alfalfa leaves at a concentration of about 14 μg/ml. This compares favourably with the concentration which has a marked effect on germ-tube growth of the fungus. The identification of medicarpin followed extensive work on the accumulation of a phytoalexin in infection droplets on alfalfa in response to infection by non-pathogens (Higgins and Millar, 1968), and on the ability of some pathogens of alfalfa to degrade the phytoalexin *in vitro* (Higgins and Millar, 1969a,b, 1970). *Stemphylium botryosum*, a pathogen of alfalfa, destroyed antifungal activity; *S. loti*, a weak pathogen, converted the phytoalexin to another active compound; *Colletotrichum phomoides* did something similar; and the other non-pathogen used, *H. turcicum*, failed to degrade the phytoalexin. Medicarpin is thus a compound of great interest, but it remains desirable to know the types of change in medicarpin concentration which follow infection inside tissues of alfalfa.

Quite a different sort of substance, safynol (8) a polyacetylene (Thomas and Allen, 1970), has been isolated from safflower, *Carthamus tinctorius*. It has been extracted from the hypocotyls in the following amounts: from un-wounded hypocotyls, 0·55 μg/g fresh weight; wounding caused a slight increase in the content of safynol (to 1·2 μg/g). After infection by the

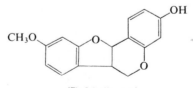

(7) Medicarpin

(8) Safynol

pathogen *Phytophthora dreschleri* for 4 days, between 10 and 11 μg/g were obtained. The ED_{50} of safynol to mycelial growth was 12 μg/ml, and it completely prevented mycelial growth at a concentration of 30 μg/ml. Thus, safynol was effective, although not completely, against the infecting agent at the concentrations in which it was produced.

At the present time therefore, a wide range of chemical substances have been isolated which behave, in a general sense, as phytoalexins.

V. PHYTOALEXINS AND DISEASE RESISTANCE IN BEANS
(*Vicia* AND *Phaseolus*)

For most of the rest of this review, I wish to consider our experiences with phytoalexins, working mainly with two diseases. The first, a rather simple one, is the Chocolate Spot disease of broad bean, *Vicia faba*. Leaves of *V. faba* respond to penetration by *Botrytis* by undergoing necrosis and browning so that a brown leaf spot is the immediate response to infection by this fungus. *Botrytis cinerea* is usually limited to initial infection sites, whereas *B. fabae* has the capacity, under some circumstances, to spread rapidly through the leaf from these infection sites. Our basic interest was to understand the nature of the processes that controlled the limitation of *B. cinerea* and the ability of *B. fabae* to spread through the tissue. The second disease with which we have worked is Anthracnose of bean caused by the fungus *Colletotrichum lindemuthianum*. *Colletotrichum* is a particularly interesting organism because, although it can be cultured *in vitro* quite readily and therefore behaves in a similar way to the facultative parasites, it exhibits many of the features of the obligate parasites. First, it exists as a number of physiological races; each one is specifically able to attack particular varieties of bean only. In susceptible varieties a compatible race can grow in apparent harmony with host cells for many days following infection, whereas in resistant varieties the immediate response of the host cells to attempted infection is a hypersensitive reaction; the fungus is then limited to the hypersensitive cells. The existence of the fungus in harmony with the susceptible protoplast for a number of days is the important feature which makes *Colletotrichum* appear to resemble the obligate parasites such as the rusts in their host/parasite relationships.

A. BROAD BEAN, *Botrytis*, WYERONE ACID AND WYERONE

We first started to work with the Chocolate Spot disease of *V. faba* before we had any particular interest in phytoalexins (Deverall and Wood, 1961a,b) and it was our search for the nature of the process causing the limitation of *B. cinerea* that eventually led us to the idea that a phytoalexin might be involved. Many experiments suggested that *B. cinerea* was being restricted to infection sites by a process of inhibition which probably developed after infection (Purkayastha and Deverall, 1965a). We (Purkayastha and Deverall,

9

1965b; Deverall, 1967) adopted the technique of Müller (1958) of incubating spores in seed cavities of pods, and I shall describe typical results after incubating 5×10^5 spores per ml of either *B. cinerea* or *B. fabae* for different periods of time in the pod cavities of *V. faba*. Infection droplets containing *B. cinerea* became highly antifungal between 9 and 18 hours after placing them in seed cavities and were completely fungistatic towards *B. cinerea* after incubation for 24 hours, whereas those droplets containing *B. fabae* were highly stimulatory to fungal growth even after incubation for 24 hours in seed cavities.

A variety of unpublished experiments were performed by me and my colleagues to discover whether the detected activity was coming from the fungus or from the pod. First, we tested substances emanating from the fungus. For example, spores of *B. cinerea* shaken in water at a concentration of 10^7 per ml for 24 hours did not germinate, but the fluid which had bathed these spores proved to be slightly stimulatory to the germination of test spores. The same result was obtained using a concentration of 10^6 spores per ml in water, which in shaken suspension did germinate. Spores germinated in 1 % glucose for 24 hours provided a markedly stimulatory bathing solution to test spores. Complete culture medium after mycelial growth for 7 days was also stimulatory to the germination and germ-tube growth of test spores. Thus there was no indication of the formation of an extractable antifungal substance arising from the fungus. When water was incubated in seed cavities of pods, it was highly stimulatory to fungal growth, but cell-free culture filtrates, obtained as above, were completely inhibitory to germination of test spores after incubation in pods for 24 hours. This suggested that substances in the culture filtrates caused the pods to make an antifungal material. In passing, I should mention that the culture filtrates caused browning of the host tissues beneath the droplets. Ultraviolet spectrophotometry of ether-soluble materials obtained from inhibitory diffusates from pods into culture filtrates showed that a characteristic absorption spectrum was associated with the active principle; this spectrum was lacking in extracts from the other types of treatment.

$$CH_3CH_2CH{=}CHC{\equiv}CCOC{=}CHCH{=}CCH{=}CHCOOR$$
$$\lfloor{-}\!\!-O\!\!-{-}\rfloor$$

(9) Wyerone, R = CH_3;   (10) Wyerone acid, R = H

The antifungal compound obtained from pods always behaved as a single substance in different chromatographic and solvent separation procedures (Deverall, 1967). At Wye College, an antifungal substance called wyerone (9) an acetylenic-keto-methyl-ester, was isolated from healthy broad bean tissues (Fawcett *et al.*, 1968, 1969) and workers at Imperial College were able to identify the phytoalexin in infected tissues as wyerone acid (10), the free acid of which wyerone is the methyl ester (Letcher *et al.*, 1970). The ED_{50} of wyerone acid against germination of *B. cinerea* was 9 μg/ml and against that of *B. fabae*

26 μg/ml, indicating a lower sensitivity of the more pathogenic fungus to wyerone acid. Eighteen μg/ml of wyerone acid prevented spore germination of *B. cinerea* whereas 45 μg/ml achieved the same against *B. fabae*. Yields of wyerone acid from different types of tissues were as follows (Letcher *et al.*, 1970). Infection droplets containing *B. cinerea* in pods yielded between 9 and 30 μg wyerone acid/ml which equal or exceed the values likely to have a 50% or completely inhibitory effect on spores of *B. cinerea*. Therefore the activities of crude diffusates can be accounted for completely by the yields of wyerone acid obtained from diffusates. Traces of wyerone acid were found in infection droplets containing *B. fabae*; the levels found were too low to have any effect on *B. cinerea*, so again this is in accord with the bioassay results on crude diffusates. Quite different results were obtained when whole leaves bearing scattered lesions were extracted. Markedly higher yields of wyerone acid were obtained from leaves bearing lesions caused by *B. fabae* than those caused by *B. cinerea*, in accord with results published before the phytoalexin was identified (Deverall *et al.*, 1968). This major difference between results obtained using pods and leaves led to a new series of investigations on events in leaves (Deverall and Vessey, 1969), in terms of bioassays of extracts from different parts of leaves. Healthy tissue yielded no inhibitor. The flecked brown tissue immediately beneath infection droplets containing *B. cinerea* was moderately inhibitory towards germination of test spores, the black tissue interwoven with hyphae beneath droplets containing *B. fabae* 2 days after infection was non-inhibitory towards the fungus but the peripheral zone of apparently healthy tissue around black lesions was completely inhibitory to germination in bioassay. Recent research by Mansfield shows that wyerone acid is formed in or around damaged cells in *V. faba* irrespective of the infecting *Botrytis* species and that wyerone acid is metabolized *in vitro* and *in vivo* particularly readily by the relatively insensitive fungus *B. fabae*.

My present view is that in our experiments sufficient wyerone acid accumulates around damaged cells at multi-spore infection sites to prevent further growth of *B. cinerea* under most circumstances and to slow the growth of *B. fabae*. However *B. fabae* can often degrade the wyerone acid and continue to spread through the leaf. These ideas may be tested by examining the effects of other treatments which change the normal development of the fungi, and they are supported in part by a study of the action of pollen grains in breaking resistance to *B. cinerea*. Substances emanating from pollen do not prevent accumulation of wyerone acid, but they render *B. cinerea* relatively insensitive to the action of wyerone acid (Mansfield and Deverall, 1971).

The demonstration of wyerone acid as a phytoalexin in broad bean prompted research at Wye College on changes in wyerone levels in *V. faba* and it is of great interest that substantial increases in levels of wyerone followed infection by *B. fabae* (Fawcett *et al.*, 1971). It remains to be revealed whether syntheses of both wyerone and wyerone acid are promoted by cellular damage after infection, or whether synthesis of one of these compounds precedes conversion to the other by the plant or fungus in the leaf. Wyerone might be formed in other types of infection to act as a phytoalexin against certain pathogens; it seems unlikely that it is effectively antifungal *in vivo* against *Botrytis*, having an ED_{50} of near 100 μg/ml against this fungus (Fawcett *et al.*, 1969).

B. BEAN, *Colletotrichum* AND PHASEOLLIN

We have also attempted to evaluate the role of phytoalexins in resistance to the Anthracnose disease of beans caused by *Colletotrichum*. Using infection droplets in pods of different varieties of bean, no relationship was found between high antifungal activity generated within 24 hours and the resistance or susceptibility of the varieties (Deverall *et al.*, 1968). In some varieties, phaseollin may have contributed to the antifungal activity, but another compound, possibly that detected after *Rhizoctonia* infections of bean (Pierre and Bateman, 1967), was also prominent (Smith, 1970). Since the interiors of pods may be quite unlike the outer surfaces of stems and hypocotyls as sites for attempted infection, a new series of experiments was started to study phaseollin changes inside bean hypocotyls as they underwent interactions with different physiological races of *C. lindemuthianum*. Biological investigations by Skipp (1971) confirmed the work of Rahe *et al.* (1969) and earlier workers, that in a resistant variety of bean an immediate response to penetration, which occurred about 2 days after infection, was death and browning of the host cell and failure of the fungal germ-tube to develop further beneath the spherical appressorium on the surface of the cell. In the susceptible plant the germ-tube emerged into the cell from beneath the appressorium and grew readily between the cell wall and the protoplast without causing any apparent damage to the host cell. Hyphal growth continued for several days and broad hyphae passed from one cell to the next without signs of visible damage until about 6 days after infection when necrosis, browning and tissue collapse occurred quite suddenly. It is this sudden collapse and browning of the tissue which gives rise to the characteristic symptoms of the Anthracnose disease. Dr Bailey, working closely with Dr Skipp at Imperial College, measured changes in phaseollin concentration in bean hypocotyls following infection. The timing of the rapid formation of phaseollin differed greatly in resistant and susceptible cells. The period of rapid formation of phaseollin in a resistant plant started on the second day and coincided with the appearance of the hypersensitive browning of the cells. In the susceptible plant there was a negligible change in the low background level of phaseollin throughout the long period of mycelial growth in the tissue, and phaseollin was not formed rapidly until the stage when the brown sunken anthracnose lesions were produced, thus confirming the results of Rahe *et al.* (1969). The concentrations of phaseollin in small excised pieces of tissue bearing the hypersensitive flecks or the dark sunken lesions typical of the susceptible reaction were between 70–100 μg/g fresh weight of tissue, negligible amounts being present in neighbouring green tissue (Bailey and Deverall, 1971). The effect of purified phaseollin on the germination, germ-tube growth and mycelial growth of *C. lindemuthianum* was also studied. With respect to the effects of phaseollin on germination, all races were approximately equally sensitive to phaseollin; between 5 and 10 μg/ml phaseollin solubilized in different agents prevented germination. Application of about 3 μg/ml of phaseollin to existing germ-tubes caused them to shrink and abort. On the other hand, mycelium in liquid culture of *Colletotrichum* was much less sensitive to phaseollin, since it often tolerated and apparently metabolized concentrations of phaseollin of 50 μg/ml. The important point was that the concentrations of phaseollin detected in excised hypersensitive tissue, which contained a lot of healthy unaffected cells as well as brown cells, effectively prevented the growth and development of germ-tubes of all races of *C. lindemuthianum*.

This research makes it seem very likely that phaseollin and other unidentified compounds are responsible for the cessation of growth of germ-tubes into hypersensitive cells in bean but a remaining question is whether development of germ-tubes changes in the hours before or after phaseollin begins to accumulate. Specificity in the Anthracnose disease may depend upon the ability of particular races either to avoid provoking or to suppress processes leading to cellular damage, and thence phaseollin synthesis, in susceptible varieties of bean.

VI. PHYTOALEXINS AND BACTERIAL DISEASES

The observation that phaseollin formation is closely associated with cellular browning in bean is similar to that of Rahe *et al.* (1969) and suggests the possibility that phaseollin might be formed in bean leaves after other types of infection causing cellular browning. We therefore investigated the effect of the bacterium *Pseudomonas phaseolicola*, the cause of halo blight in bean, on phaseollin formation. This bacterium exists as at least two physiological races. In the variety Red Mexican, race 1 causes a hypersensitive browning of the cells within 3 days of infection, whereas race 2 does not cause browning and leads to the development of the characteristic halo blight symptoms. When hypersensitivity was observed, phaseollin was found in leaves in concentrations of the order of 200 μg/g fresh weight. Some phaseollin (20 μg/g) was detected in infected susceptible tissue but no phaseollin could be detected in healthy leaves (Stholasuta *et al.*, 1971). This was a clear indication that cellular necrosis, whether caused by a fungus or by a bacterium, was linked to the formation of phaseollin and there was no specific need for fungal infection in phaseollin formation. However when the bacterium *Ps. mors-prunorum*, which also causes an hypersensitive response in bean, was used, it was not possible to obtain phaseollin after hypersensitivity to this organism. Obviously there is much to be learned about the control of formation of phaseollin after infection. Recent work has shown that the virus TNV, as well as incompatible races of rust fungi which cause necrosis in bean, also induces the formation of high levels of phaseollin at the time when browning is observed (Bailey and Ingham, 1971).

Demonstrations that a bacterium, and particularly a virus, can cause phaseollin formation after inducing cellular damage make it seem unlikely that phaseollin formation is under the control of specific pathogen-secreted inducing substances such as monilicolin A (Cruickshank and Perrin, 1968) during infection. It seems much more likely that phaseollin formation is a non-specific consequence of cellular damage caused by incompatible organisms and substances. Quite possibly, certain organisms such as *Ps. mors-prunorum* may be able to so affect particular pathways of aromatic biosynthesis that phaseollin does not accumulate, but this is a subject requiring more research.

Brief mention will now be made of two other recent discoveries relating to antibacterial substances produced in plants in response to infection by bacteria. Stall and Cook (1968) showed that in a pepper variety which undergoes an hypersensitive reaction to the bacterium *Xanthomonas vesicatoria* it was possible to obtain, by washing out intercellular fluids, a high yield of unidentified antibacterial substance(s) 16 hours after inoculation with the bacterium. Similar activity was found 84 hours after inoculation of susceptible tissue. Lozano and Sequeira (1970) showed that when hypersensitive reactions occurred in tobacco leaves to the bacterium *Ps. solanacearum* it was possible, using a similar technique, to detect low yields of antibacterial activity 12 hours after inoculation and high yields 20 hours after inoculation. The substances responsible for this antibacterial effect disappeared at later stages. Again, no information is available about the nature and importance of the antibacterial substance(s) but these reports, coupled with the demonstration that bacterial infection can cause phaseollin formation, should stimulate investigations on the role of phytoalexins in resistance to bacterial diseases of plants.

VII. GENERAL CONSIDERATIONS

I should now like to present my working hypothesis concerning defence mechanisms which are activated after infection of plants. First, plants make antimicrobial compounds in response to cellular damage. Evidence for this, with respect to damage resulting from infections, has been presented above and I have made reference to facts that certain forms of physical and chemical damage can cause some of these antimicrobial compounds to form, mechanical wounding being sufficient in some cases. Wyerone acid, for example, can be formed by an appropriate form of bruising of the bean leaf, and chemical toxicants cause high levels of pisatin to form in pea, of ipomeamarone in sweet potato and of orchinol in orchid tubers. Secondly, different types of compound are made in different plants. Abundant evidence has been presented for this in this review. *The following two points are of particular importance.* Thirdly, I suggest that some microorganisms are successful parasites because they overcome the defensive response which they activate on damaging host cells, having a low sensitivity to and/or a capacity to degrade the compounds produced. As an example of this, I suggest *B. fabae* in broad bean which has a lower sensitivity to wyerone acid than *B. cinerea* and a higher capacity to degrade wyerone acid. Published evidence has been referred to of a similar type of phenomenon with respect to medicarpin in alfalfa, and indications have been quoted that some of the successful parasites of pea which cause damage after infection degrade pisatin. Fourthly, I suggest that some other microorganisms are successful parasites because they avoid causing immediate cellular damage in their hosts and thus avoid provoking the defensive response. Typical of this phenomenon are *C. lindemuthianum*, just discussed, and perhaps *P. infestans* as revealed by the work of Tomiyama *et al.* (1968) on rishitin. A further elaboration of this fourth point is that compatible races of *P. infestans* may suppress processes leading to cellular damage and rishitin formation, as suggested by Varns and Kuć (1971). Prior inoculation with a compatible race were found to diminish both responses when discs were later inoculated with an incompatible race. Some form of depletion of host metabolites by growth of compatible mycelium in susceptible cells in a disc, which is a limited body of tissue, could prevent the same or neighbouring cells from making rapid and extensive responses to incompatible germ-tubes. Therefore this interesting demonstration by Varns and Kuć does not necessarily imply that compatible germ-tubes specifically suppress attempts by susceptible cells to undergo hypersensitivity following infection. However it seems clear that research is needed on the events which occur in susceptible cells when they are colonized apparently harmoniously by compatible pathogens, such as appropriate races of some bacteria and fungi including *Colletotrichum*, *Phytophthora* and the rusts.

Finally, I should like to give my present view of phytoalexins. Now the term phytoalexin is very convenient, acting as a code-word to indicate that one is considering substances formed in effective amounts in infected plants. Phytoalexins, as they have been revealed, are antimicrobial metabolites present in low amounts in healthy plants and accumulating in high concentrations in or around cells damaged by many different stimuli; I suggest that their accumulation is part of a general repair and defence system triggered off by damage, particularly the relatively continuous form of damage during pathogenesis by incompatible organisms. This would imply that phytoalexins cannot be considered as the only substances formed specifically in response to certain sorts of infection but that they are the antimicrobial components of a range of compounds produced by plant cells in response to damage.

REFERENCES

Bailey, J. A. (1969). *Phytochemistry* **8**, 1393–1395.
Bailey, J. A. (1970). *J. gen. Microbiol.* **61**, 409–415.
Bailey, J. A. and Deverall, B. J. (1971). *Physiol. Plant Path.* **1**, 435–449.

232   B. J. DEVERALL

Bailey, J. A. and Ingham, J. L. (1971). *Physiol. Plant Path.* 1, 451–456.
Bernard, N. (1909). *Ann. Sci. nat. Bot.* 9, 1–196.
Bernard, N. (1911). *Ann. Sci. nat. Bot.* 14, 221–234.
Chester, K. S. (1933). *Quart. Rev. Biol.* 8, 129–154 and 275–324.
Christenson, J. A. (1969). *Phytopathology* 59, 10.
Condon, P. and Kuć, J. (1960). *Phytopathology* 50, 267–270.
Condon, P. and Kuć, J. (1962). *Phytopathology* 52, 182–183.
Cruickshank, I. A. M. and Perrin, D. R. (1960). *Nature, Lond.* 187, 799–800.
Cruickshank, I. A. M. (1962). *Aust. J. biol. Sci.* 15, 147–159.
Cruickshank, I. A. M. (1963). *A. Rev. Phytopathol.* 1, 351–374.
Cruickshank, I. A. M. and Perrin, D. R. (1963a). *Life Sci.* 2, 680–682.
Cruickshank, I. A. M. and Perrin, D. R. (1963b). *Aust. J. biol. Sci.* 16, 111–128.
Cruickshank, I. A. M. and Perrin, D. R. (1968). *Life Sci.* 7, 449–458.
Cruickshank, I. A. M. and Perrin, D. R. (1971). *Phytopath. Z.* 70, 209–229.
Deverall, B. J. (1967). *Ann. appl. Biol.* 59, 375–387.
Deverall, B. J. and Wood, R. K. S. (1961a). *Ann. appl. Biol.* 49, 461–472.
Deverall, B. J. and Wood, R. K. S. (1961b). *Ann. appl. Biol.* 49, 473–487.
Deverall, B. J., Smith, I. M. and Makris, S. (1968). *Neth. J. Plant Path.* 74, 137–148.
Deverall, B. J. and Vessey, J. C. (1969). *Ann. appl. Biol.* 63, 449–458.
De Wit-Elshove, A. (1968). *Neth. J. Plant. Path.* 74, 44–47.
De Wit-Elshove, A. (1969). *Neth. J. Plant Path.* 75, 164–168.
De Wit-Elshove, A. and Fuchs, A. (1971). *Physiol. Plant Pathol.* 1, 17–24.
Fawcett, C. H., Spencer, D. M., Wain, R. L., Fallis, A. G., Jones, E. R. H., Le Quan, M., Page, C. B., Thaller, V., Shubrook, D. C. and Whitham, P. M. (1968). *J. chem. Soc.* (C) 1968, 2455–2462.
Fawcett, C. H., Spencer, D. M. and Wain, R. L. (1969). *Neth. J. Plant Path.* 75, 72–81.
Fawcett, C. H., Firn, R. D. and Spencer, D. M. (1971). *Physiol. Plant Pathol.* 1, 163–166.
Gaümann, E. (1964). *Phytopath. Z.* 49, 211–232.
Gaümann, E. and Jaag, O. (1945). *Experientia* 1, 21–22.
Gaümann, E. and Kern, H. (1959a). *Phytopath. Z.* 35, 347–356.
Gaümann, E. and Kern, H. (1959b). *Phytopath. Z.* 36, 1–26.
Gaümann, E. and Hohl, H. R. (1960). *Phytopath. Z.* 38, 93–104.
Gaümann, E., Nüesch, J. and Rimpau, R. H. (1960). *Phytopath. Z.* 38, 274–308.
Hardegger, E., Biland, H. R. and Corrodi, H. (1963). *Helv. Chim. Acta* 46, 1354.
Higgins, V. J. and Millar, R. L. (1968). *Phytopathology* 58, 1377–1383.
Higgins, V. J. and Millar, R. L. (1969a). *Phytopathology* 59, 1493–1499.
Higgins, V. J. and Millar, R. L. (1969b). *Phytopathology* 59, 1500–1506.
Higgins, V. J. and Millar, R. L. (1970). *Phytopathology* 60, 269–271.
Hiura, M. (1943). *Sci. Report Gifu Agric. Coll. Japan* 50, 1–5.
Katsui, N., Murai, A., Takasugi, M., Imaizumi, K., Masamune, T. and Tomiyama, K. (1968). *Chem. Commun.* 1968, 43–44.
Kubota, T. and Matsuura, T. (1953). *J. chem. Soc. Japan, Pure Chem. Sect.* 74, 248–251.
Kuć, J. (1968). *World Rev. Pest Control* 7, 42–55.
Letcher, R. M., Widdowson, D. A., Deverall, B. J. and Mansfield, J. W. (1970). *Phytochemistry* 9, 249–252.
Lozano, J. C. and Sequeira, L. (1970). *Phytopathology* 60, 833–838.
Magrou, J. (1924). *Rev. Path. Vég. Entom. Agr.* 11, 189–192.
Mansfield, J. W. and Deverall, B. J. (1971). *Nature, Lond.* 232, 339.
Müller, K. O. (1958). *Aust. J. biol. Sci.* 11, 275–300.
Müller, K. O. and Börger, H. (1941). *Arb. biol. Anst. (Reichsanst) Berl.* 23, 189–231.
Nobécourt, P. (1923). *C. R. Acad. Sci. Paris* 177, 1055–1057.

Perrin, D. R. (1964). *Tetrahedron Letters* 29–35.
Perrin, D. R. and Bottomley, W. (1962). *J. Am. chem. Soc.* **84**, 1919–1922.
Perrin, D. R. and Cruickshank, I. A. M. (1965). *Aust. J. biol. Sci.* **18**, 803–816.
Pierre, R. E. and Bateman, D. F. (1967). *Phytopathology* **57**, 1154–1160.
Purkayastha, R. P. and Deverall, B. J. (1965a). *Ann. appl. Biol.* **56**, 139–147.
Purkayastha, R. P. and Deverall, B. J. (1965b). *Ann. appl. Biol.* **56**, 269–277.
Rahe, J. E., Kuć, J., Chuang, C.-M. and Williams, E. B. (1969). *Neth. J. Plant Path.* **75**, 58–71.
Schwochau, M. E. and Hadwiger, L. A. (1968). *Archs Biochem. Biophys.* **126**, 731–733.
Skipp, R. A. (1971). Ph.D. thesis. University of London.
Smith, D. G., McInnes, A. G., Higgins, V. J. and Millar, R. L. (1971). *Physiol. Plant Pathol.* **1**, 41–44.
Smith, I. M. (1970). *Ann. appl. Biol.* **65**, 93–103.
Smith, I. M. (1971). *Physiol. Plant Pathol.* **1**, 85–94.
Sondheimer, E. (1961). *Phytopathology* **51**, 71–72.
Stakman, E. C. (1915). *J. agric. Res.* **4**, 193–199.
Stall, R. E. and Cook, A. A. (1968). *Phytopathology* **58**, 1584–1587.
Stholasuta, P., Bailey, J. A., Severin, V. and Deverall, B. J. (1971). *Physiol. Plant Pathol.* **1**, 177–184.
Thomas, C. A. and Allen, E. H. (1970). *Phytopathology* **60**, 261–263.
Tomiyama, K., Sakuma, T., Ishizaka, N., Sato, N., Katsui, N., Takasugi, M. and Masamune, T. (1968). *Phytopathology* **58**, 115–116.
Uritani, I. and Akazawa, T. (1955). *Science, N.Y.* **121**, 216–217.
Uritani, I., Uritani, M. and Yamada, H. (1960). *Phytopathology* **50**, 30–34.
Varns, J. L. and Kuć, J. (1971). *Phytopathology* **61**, 178–181.
Varns, J. L., Kuć, J. and Williams, E. B. (1971). *Phytopathology* **61**, 174–177.
Ward, H. M. (1905). *Ann. Bot. (Lond.)* **19**, 1–54.

CHAPTER 14

Orobanche and Other Plant Parasite Factors

W. G. H. EDWARDS

Chemistry Department, Royal University of Malta, Malta

I. INTRODUCTION

Many higher plants are parasitic on others, attaching themselves to stems (for example *Cuscuta*) or to roots, and among the latter are numerous species of *Striga* (witchweed) and *Orobanche* (broom-rapes). Although these latter genera of angiosperms are of considerable (and increasing) importance as weeds of economic crops, they are also, in their own right, of fundamental biological and chemical interest. In the case of *Orobanche*, the spectacle of clumps of white, yellow or purple flowers arising from the base of cultivated plants, so that the farmer reaps where he has not sown, as it were, has seemed almost magical to the peasant since the earliest recorded times. Wilhelm (1962), in a review of the history of *Orobanche*, has presented evidence of its identification as a parasite as far back as the Grecian period, although serious botanical study of it dates from around the 16th century. The voluminous literature on *Striga* and *Orobanche* need not be reviewed fully here, as a number of good general reviews already exist (Krenner, 1958; Privat, 1960; Parker, 1965; Williams, 1961) as well as several selected Bibliographies published by the Weed Research Organisation of the Agricultural Research Council. The emphasis of these papers is on the botanical (e.g., Kadry and Tewfik, 1956) and agricultural aspects of the parasites. The fact that chemical factors are involved in the germination of their seeds and their relationship to the host plants has come to be recognized since the middle of the nineteenth century, and it is this aspect of the subject that is reviewed here. Systematic studies of chemical factors go back about 35 years, and the earlier work has been well summarized by

Brown (Brown, 1946; Brown *et al.*, 1951a) and by Sunderland (1960a,b). Indeed, the work of these two, at Leeds University, and their collaborators at Cambridge, notably A. W. Johnson and A. R. Todd (see Brown *et al.*, 1949c, 1952) have formed the basis of all approaches made to the problem since that time.

It must be recognized at the beginning that the study of the chemical factors involved in the growth of these root parasites has scarcely begun. The central chemical problem is the nature of the stimulant which triggers germination of the seed, but possibly connected with it is the agricultural problem of controlling these weeds. Wilson Jones (1953) has pointed out that in North Africa poor crops have often been attributed to exhausted soil and inefficient farming methods, when a good deal of the blame should be laid against parasitism by *Striga*. In Malta and in Sicily there has been a steady decline in production of leguminous crops such as peas and beans, attributable to the heavy losses resulting from *Orobanche* infestation. Of the sixty or so species of *Striga* and ninety or so species of *Orobanche* (Beck-Managetta, 1930), many are crop parasites (see Table I). Although *Striga* species are chlorophyll-containing, and therefore not entirely parasitic, they do in fact withdraw a good deal of their nutrients from the host plants, and *Orobanche* species are entirely parasitic.

TABLE I

Common species of *Striga* and *Orobanche*

Name	Typical hosts	Region
Striga		
S. hermonthica Benth.	Sorghum, maize	Africa
S. lutea Lour (=*S. asiatica* Kuntze)	Rice, maize	Africa, Asia, U.S.A.
S. orobanchoides Benth.	Graminae generally, also tobacco	Africa, Asia
Orobanche		
O. crenata Forsk.	Beans and other legumes	Mediterranean, Egypt
O. hederae Duby	Ivy	Europe
O. ramosa L.	Tomato, tobacco, lettuce	U.S.A., Italy
O. minor Sm.	Clover	Europe, Australia
O. ludoviciana Nutt.	Tomato	U.S.A.
O. cernua Loefl. (syn. *O. cumana* Wallr.)	Brassica, tobacco, sunflower	India, U.S.S.R.

Numerous methods of control of these parasites in the field have been, and are being, studied. These include applications of herbicides (Puzzilli, 1969; Netzer, 1969; Atanas, 1969) as well as soil fumigants such as methyl bromide

(Netzer, 1961; Wilhelm *et al.*, 1959), rindite (Izard and Hitier, 1958), and dibromochloropropane (Zahran, 1970).

These techniques have so far met with indifferent or variable success, and are probably unable at the present time to arrest the spreading of infection. The seeds of *Striga* and *Orobanche* are extremely small (of the order of 0·25 mm diameter) and easily capable of transportation by wind, birds, seeds of crop plants (possibly *Striga hermonthica*, first reported in the U.S.A. in 1956, was carried in this way), farm machinery, flooding, and movement of soil (as in Malta today). When one considers that about 50,000 seeds per flower spike are produced by *Orobanche minor*, which is typical of this genus, and that seed remains viable in the soil for 10 to 15 years, it is clear that only an extremely efficient and convenient eradication method is likely to be effective.

It is for this reason that several workers have directed their attention to investigating the chemical stimulants produced not only by host plant roots, but by the roots of others which are not themselves parasitized (Muller, 1941). If it were possible to germinate dormant seeds in the soil in the absence of the host plant, or to prevent the chemical stimulant from effecting germination, the problem would be within sight of solution.

II. Germination Processes

Any review of the chemical factors involved in parasitism by *Orobanche* and *Striga*, and similar plants, must recognize three stages at which such factors are, or may be, operative. These are the state of dormancy, the initiation of germination, and the post-germination development of the seed.

A. DORMANCY

As is common with seeds, a period of pre-conditioning is necessary before seeds of *Orobanche, Striga, Alectra vogeli* (Botha, 1951) and *Cuscuta* (Fritsche *et al.*, 1958) parasites are capable of germinating. This involves a period of exposure to water, which ranges from a few days in the case of *Alectra* to about 3 weeks for *S. hermonthica*. In Fig. 1 is shown the variation in the maximum percentage of germination of seeds of two *Orobanche* and one *Striga* species, when exposed to water at 23–25°C. It will be noted that in the case of *Striga*, the period over which conditioned seed remains germinable is short, compared with that for *Orobanche*. However, in our own experiments we have found that *O. crenata* does slowly fall in germinating power with time, and there is normally a useful maximum period of about 2 months. The difference is, therefore, one of degree.

In spite of this pre-treatment, the percentage of seeds which will germinate when exposed to stimulant from host root or a root exudate varies appreciably from sample to sample, and from species to species. Thus, Brown and his coworkers report, for *S. hermonthica*, germination percentages from 10 to 60, and Worsham (Worsham *et al.*, 1964) finds 90% with *S. lutea*. Among various *Orobanche* species, values of 5% (Rao, 1955 for *O. cernua*), 10–60% (Nash and Wilhelm, 1960; Racovitza, 1958; Izard and Hitier, 1954 for *O. ramosa*), and 70–90% (Cappelletti, 1936; Chabrolin, 1938 for *O. crenata*) have been obtained. Although these variations may occasionally be due to low essential viability resulting from absence of pollinating insects (Nash and Wilhelm, 1960), there is no doubt that a period of after-ripening is commonly required, and that the embryo of the freshly dehisced seed is immature (Privat, 1960). The period of after-ripening may extend over years. Samples of seed collected in May 1968 in Malta, and stored at room temperature, give a percentage germination today of 45%, whereas the corresponding figures for seed collected in May 1969 and May 1970 are 9% and zero respectively. As is not uncommon, storage at low temperatures often improves germination, as in the

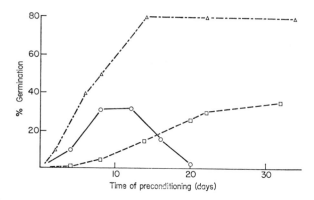

FIG. 1. Effect of water preconditioning on % germination of parasite seeds. △— · —△ *O. minor*, o——o *S. hermonthica* (Brown *et al.*, 1949a, 1951a) □– – –□ *O. crenata* (Malta, 1970)

case of *Cuscuta* (Fritsche *et al.*, 1958). In *O. crenata*, the proportion of germinable seeds nearly trebled after a 6-week storage at 5°C.

What takes place during water pre-treatment of the parasite seeds is not known. Many workers have considered that chemical dormancy factors are dissolved out of the seed, and certainly soluble compounds rapidly appear in the water. Privat (1960) has suggested that, since some *Orobanche* species contain chlorogenic acid and a caffeic acid glycoside (Bridel and Charaux, 1924), these acids and other phenolics might be responsible for the dormancy of *Orobanche* seed, and that the function of the water treatment is primarily to leach out these inhibitors. He was not able, however, to detect phenols in the leachings from *O. hederae*. In the case of *O. crenata*, we have found it simple to show the presence of inhibitors in the seed, as in Fig. 2, and to make it appear that Privat's suggestion may be in part a correct one.

It had already been suggested by Brown (1946) that during the water pre-treatment the seed itself produces a stimulant similar to that released by the host plant roots, on the general grounds that the response of the seed to a host stimulant solution increases with increase of the period of pre-treatment, at least over a certain part of the process. In the case of *Alectra vogelii*, Botha (1951) has confirmed this observation, but in the light of the existence of endogenous inhibitors the theory needs modification, perhaps to involve competition between residual inhibitor in the "conditioned" seed and a combination of endogenous and host stimulants (Villiers *et al.*, 1963). Certainly the evidence we have is against the appearance of any great amount of endogenous stimulant. This would lead us to expect more spontaneous germination of well-conditioned seed, in which inhibitor concentration is low, than is actually the case. Most workers have observed some spontaneous germination, usually much less than 1 %, both with *Striga* and *Orobanche*, but there are two recorded

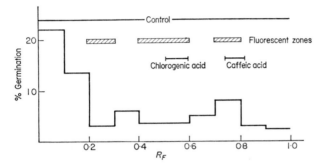

FIG. 2. Germination inhibitors in *O. crenata* seeds. 5 mg of seeds extract chromatographed on paper using *t*-butanol/ethanol/water (4:2:2). Fractions tested in presence of purified *O. crenata*/*V. faba* factor

instances of much higher values. Nash and Wilhelm (1960) found 20% of spontaneous germination with *O. ludoviciana*, and Krenner (1958) reports 18–21% spontaneous germination of a seed sample of *O. cumana* when tested in the same year as it was collected, a similar observation on freshly gathered *Cuscuta* seed having been made by Degen (1911). With *O. crenata* we only rarely obtain any germination of seed in the absence of the host stimulant. The observation of Nash and Wilhelm must be considered in the light of the remarkable stimulation of germination of *O. ludoviciana* seed by gibberellic acid, although other workers (e.g. Cook *et al.*, 1966, and the Malta group) have not found this effective in other species. There is clearly a species difference operative here, and one can only conclude that in general the seeds of *Orobanche* and *Striga* do require the application of the external host stimulant in order to germinate.

B. GERMINATION STIMULATION

There is now no doubt of the truth of early observations (Heinricher, 1898; Pearson, 1912) that in the exudates of roots of many plant species, both hosts and non-hosts, there are chemical factors that "trigger" the germination of seeds of root parasites. It is these specific substances that have come to be known, for example, as *Orobanche* and *Striga* "factors", although from what has been stated above it is clear that there are other chemical factors involved in the whole process of parasitism. It is also now clear that different stimulants exist in different plants, and that not all species of a given parasite genus are equally affected, or affected at all, by the exudate of a given plant. For this reason in what follows it is proposed to refer to the factors by a name composed of the two plant species involved, sometimes for convenience in non-Linnean terms. Thus the *S. lutea/Sorghum* factor is that factor exuded by sorghum roots which stimulates the germination of *Striga lutea*. Recent work in Malta and the U.S.A. as well as that of Brown and Sunderland has shown that more than one stimulant may be exuded by the roots of a given host, and it may be necessary to append a number to the name in due course.

The root exudates, which have been so frequently shown to stimulate germination, are complex mixtures of compounds. These include many amino acids (Dehay and Care, 1957), numerous carbohydrates including glucose, arabinose and possibly 2-keto-gluconic acid (1) (Brown *et al.*, 1949c), and phosphates (Brown *et al.*, 1949c). Though no phenolic compounds were detected by Brown and his coworkers (1949c) in linseed root exudates, Sunderland (1960a,b)

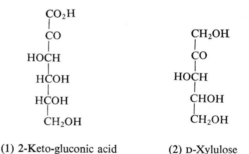

(1) 2-Keto-gluconic acid   (2) D-Xylulose

reports the presence of phenolic material in maize root exudates. None of these groups of substances appears to have a germination stimulating effect, except that proline and ornithine were reported by Donini (1959) to stimulate *O. ramosa* germination in concentrations of 10 ppm. D-Xylulose (2), which has not been certainly identified in root exudates, was shown by Brown *et al.* to have a positive though variable effect in *S. hermonthica* (1949c). Izard and Hitier (1955, 1954) reported that pyridoxin and other vitamins, possibly present in root exudates, stimulated the germination of *O. ramosa*, but this could not be confirmed by other workers (Brown *et al.*, 1951b; Nash and Wilhelm, 1960). It is certain that very small concentrations of other, highly specific, compounds are present in root exudates, which act as "triggers" for germination of these parasite seeds.

Serious attempts to isolate and characterize these specific stimulants date from 20 to 30 years ago, and relatively few groups of workers have been involved. Some indication of the difficulty of the work may be gathered from the fact that from over 300 mg of crude linseed root exudates, about 8 mg of a concentrated stimulant was obtained by Brown *et al.* (1951b) and this was later shown to contain several components, difficult to separate.

The fact that parasite seeds will only germinate within a few millimetres of the root of the stimulating plant is certainly connected with the low concentration of stimulant exuded, but also with the fact that the stimulant is rather unstable. Not only is the yield of stimulant, assessed by bioassay, rather variable (Brown *et al.*, 1949c), but it is also readily deactivated (Lansdowne, 1953; Long, 1953). Passalacqua (1943) remarked that in field trials in which moisture content and temperature were high there was less infection of beans by *O. crenata*, and many others have observed that the stimulating effect of aqueous root exudates is rapidly destroyed at elevated temperatures (Barcinsky, 1935; Chabrolin, 1937–1938; Brown, 1946). The stimulants are very sensitive to alkaline pH (Tyler, 1952, and others) and somewhat less sensitive to acid conditions above pH 4 (Brown *et al.*, 1951b). More strongly acid conditions bring about rapid deactivation; Tyler showed (1952) that a chloroform solution of *S. hermonthica*/sorghum factor lost its activity in 14 days in chloroform solution, and we have observed that the *O. crenata*/*V. faba* factor is also rapidly deactivated by chloroform at 100°C. It is probable that these effects are brought about by liberation of traces of HCl from the chloroform.

In the early work of Brown and others, in which root exudates from hydroponically-grown linseed plants were adsorbed on charcoal and later eluted, solvents containing methanol were commonly used. When purification by chromatography, or counter-current distribution techniques were subsequently employed, loss of activity was observed in many cases, and was traced to the use of primary alcohols in the solvent systems. In some of our own work on the *O. crenata*/*V. faba* factor, we have observed entire loss of activity in the removal of an active zone from a silica gel TLC plate by aqueous methanol. The effect of silica gel in promoting hydrolysis and alcoholysis of esters has been frequently observed, and these facts all point to many at least of the stimulants containing ester and/or lactone groups. In all of the factors brought to any approximate state of purity such groups have been identified, in recent years, in the IR absorption spectra (Lansdowne, 1953; Long, 1953; Tyler, 1952).

In view of these practical difficulties, it is not surprising that physical separation methods have been employed by all workers in this field. Brown and Sunderland and their coworkers used paper chromatography, with various solvents (*n*-butanol/acetic acid/water or *tert.*-butanol/ethanol/water) and were able to identify the fractions in which activity resided by bioassay, using the appropriate seeds. In all cases the main factor moved close to the solvent front (Fig. 3), with R_f values in the range 0·8–1·0, although other stimulating material was detected in some cases at R_fs in the region of 0·6. For preparative work, the

10

Cambridge and Leeds workers used column chromatography, on cellulose (Brown *et al.*, 1952) and on silica (Lansdowne, 1953). Eventually, the best concentrates were obtained after Craig partition experiments (Brown *et al.*, 1952) although it was clear that the best fractions thus obtained were not pure. It was possible, however, to come to some general conclusions as to the nature of the factors on which they worked, as a result of spectroscopic and other investigations.

The development of thin layer chromatography has much improved the

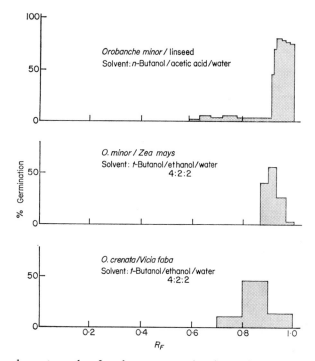

FIG. 3. Paper chromatography of crude root extracts/exudates. Linseed root exudate, tested on *O. minor* (Brown *et al.*, 1951b); *Zea mays* root exudate, tested on *O. minor* (Sunderland 1960); *Vicia faba* root extract, tested against *O. crenata* (Malta, 1970)

prospects of separation of the stimulants in a pure state. Cook and his co-workers (1966) were able to isolate in this way, from hydroponically grown cotton roots, two *S. lutea*/cotton factors. One of these was obtained crystal-line, and it was claimed to stimulate the growth of parasite seeds at concentrations of the order of 10^{-5} ppm. The other, which formed a glass and could not be crystallized, was identified as the acetate of the first compound, with an activity comparable with it, but lower. These compounds were called strigol and strigyl acetate, and on the basis of mass spectral evidence, were assigned the molecular formulae $C_{19}H_{22}O_6$ and $C_{21}H_{24}O_7$.

A review of this nature is not the place for a detailed account of our own work on the *O. crenata/V. faba* factors but it is appropriate to remark here that a small quantity of stimulant isolated last year was assigned, on mass spectral evidence, the molecular formula $C_{20}H_{24}O_6$. Separation of other factors from *V. faba*, and roots of other plants, is in progress. At the present time very little is known about the structures of the factors studied but it is clear that, as with the gibberellins, it will be necessary to depend a good deal on spectral evidence, owing to the low concentrations of stimulant actually present in root exudates. Table II lists the main infrared bands reported for several of these factors. It is clear that they have a general resemblance to one another, and that ester and/or lactone, hydroxyl, and olefinic bands are present.

TABLE II

Prominent infrared bands of stimulants from various sources

Group	*Orobanche minor/* linseed (Brown *et al.*, 1952)	*S. herm/*sorghum (Lansdowne, 1953)	*O. crenata/* *V. faba* (Malta, 1970)	*S. lutea/*cotton (Cook *et al.*, 1966)
				3590
OH	3424 broad			
			3400 broad	
C=CH				3050
				1787
				1745
C=O	1733		1725	
	1718		1720	
		1711		
				1682
	1656		1650	
			1625	
C=C		1600	1600	1601
				1550
			1510	
				1500
			1370	
C—O			1245	
	1234			

In the pioneer work of Brown and his collaborators (1949, 1951, 1952) elemental analyses of crude stimulants were carried out, and in the light of subsequent knowledge it seems likely that these preparations were too impure for the analyses to be of value. However, molecular weights were assessed as in the region of 300–350, and two hydrolysable groups were shown to be present. These are strong indications that, together with the later results referred to above, the factors are molecules of the same order of complexity as the gibberellins, one of them at least with an identical molecular formula to gibberellic acid (3).

This invites the speculation that the parasite factors are indeed gibbane derivatives. If they are, they are of a type not yet identified among the gibberellins. These latter compounds are well known as germination stimulants, for example in barley (Griffiths *et al.*, 1964), grape (Chadha and Manon, 1969), and many other seeds (Bogdanova, 1969; Anderson, 1968; Sembdner, 1963; Thompson, 1969). However, among the most potent seed-germinating gibberellins is GA_3, gibberellic acid itself (3), and this does not stimulate the

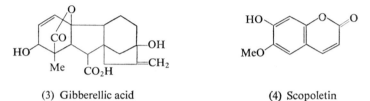

(3) Gibberellic acid   (4) Scopoletin

germination of *S. lutea* (Cook *et al.*, 1966), or of *O. crenata* in concentrations between 1 and 100 ppm, although Privat (1960) reports a small effect in concentrations of 50–100 ppm with *O. hederae*. According to Brian *et al.* (1962), some gibberellings, notably GA_2, GA_6, GA_8 and GA_9 are not active in stimulating the germination of lettuce seed, so that this negative evidence must be regarded cautiously.

The factors so far isolated are neutral compounds, although some suggestions of the presence of phenolic groups have been made (Cook *et al.*, 1966; Worsham *et al.*, 1964). These suggestions rest on colour reactions with diazonium salts, and we have also obtained evidence of the presence of phenols in the same region as active material on thin-layer plates. However, the infrared spectra do not clearly support the presence of phenol groups, and no migration of the active material is observed in electrophoresis on paper, either in *S. lutea*/ maize factor (Worsham *et al.*, 1964) or in the *O. crenata*/*V. faba* factor, in buffer solutions between pH 4·0 and 9·0. There is no doubt that phenolic compounds, in traces, do co-chromatograph with the active zone, and this may be responsible for positive colour tests reported (Sunderland, 1960a,b).

Ultraviolet absorption spectra, where reported, show the presence of bands in the region 220–325 nm. Purified *O. minor*/linseed factor absorbed at 245 nm ($\epsilon \simeq 6000$), (Brown *et al.*, 1952), and the *S. hermonthica*/sorghum factor at 253 nm ($\epsilon \simeq 9000$), (Lansdowne, 1953). With the *O. crenata*/*V. faba* factor absorption in the region 275–350 nm consisting of two or three overlapping bands, is observed, again with approximately the same molecular extinction. Cook *et al.* (1966) report the presence of a band at 236 nm ($\epsilon \simeq 18,000$) in strigol, which, coupled with the ir absorption at 1787, strongly indicates a butenolide ring as being present. None of these observations corresponds to active gibberellins, all of which lack UV absorption between 220 and 320 nm (Cross,

1954); however, the presence of absorbing impurities in the specimens studied cannot yet be firmly excluded.

Among numerous compounds tested for activity as germination stimulators for *S. lutea* (Klingman and Moreland, 1962) 4-hydroxycoumarin and 6-methoxy-7-hydroxycoumarin (4) (scopoletin) were found active in concentrations of between 2 and 20 ppm, which is of the same order as the concentrations found effective by Brown (1946) and the Malta group (Fig. 4). It was not claimed that either of these two compounds was present in natural stimulants, but Worsham *et al.* (1964) suggested that the *S. lutea*/maize factor could be a coumarin. The deactivating effect of alkali would then be ascribed to opening

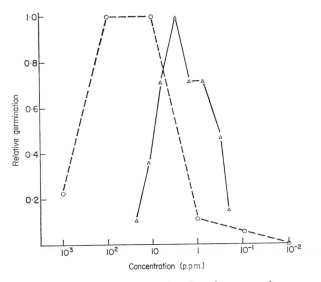

Fig. 4. Germination by stimulant solution as a function of concentration. o———o *O. minor*/ linseed factor (Brown *et al.*, 1951b) △———△ *O. crenata*/*V. faba* factor (Malta 1970)

of the lactone ring, but the fact that the activity of that factor is reported to be completely destroyed and not recoverable on acidification, does not necessarily indicate that the ring opening is irreversible. With the *O. crenata*/*V. faba* factor, solution in sodium carbonate (0·5 M) readily occurs, and the activity is recoverable on acidification, providing the solution is not left standing more than a few minutes. If solubility in this case is due to opening of a lactone ring, it is likely to be a reversible process, unless further structural changes later occur. There is no evidence that such further changes have not occurred in the alkali-deactivated *S. lutea*/maize factor.

Although it seems that the main germinating factors are lipophilic compounds, early work by Brown (1946) and by Sunderland (1960a,b) demonstrated that stimulating activity of root exudates also resides in a markedly water-soluble fraction, which appears in the carbohydrate region of the paper chromatograms. After testing a large number of compounds for stimulating activity to *Striga hermonthica*, including carbohydrates, amino-acids, organic acids, coumarins, and various unsaturated lactones, Brown and his coworkers (1949c, 1951b) concluded that D-xylulose (2) could be the germination-active pentose sugar present in these aqueous solutions. Samples of D-xylulose produced germination of conditioned *S. hermonthica* seed in concentrations of between 10^{-2} and 10^{-6} ppm. However, it was recognized that only a small part of the activity of root exudates could be ascribed to the carbohydrates, and that the compounds tested were not active to *O. minor*. After a lengthy series of studies on the effect of culturing maize roots in sugar solutions, Sunderland (1960a,b) concluded that the water-soluble stimulant could be a precursor of the *Orobanche* and *Striga* factors, and that these were both synthesized and utilized in the extending region of the host root tip. In consequence, with more mature segments of the roots, accumulation of the factors was observed, and this supported the evidence of other workers that the stimulant was mostly present in and exuded from the piliferous region of the post plant roots (Chabrolin, 1938; Rao, 1955; Barcinsky, 1934, 1935).

A previous observation (Brown *et al.*, 1949a,b) that no cell division occurred in the tip of the emerging parasite radicle, but that increasing length was due to cell extension-growth prompted these workers to study the effect of aqueous solutions of the *S. hermonthica*/sorghum factor on pea root extension. They found that at concentrations of 10 ppm, approximately a 10% increase over controls was found in extension tests with pea root segments, an effect also produced by concentrations of 0·1 ppm of D-xylulose. This cannot be taken as evidence that the parasite factor itself is a cell extension hormone, as the effect could be due to D-xylulose or some other natural hormone in the impure specimen.

C. POST-GERMINATION FACTORS

Although the actual appearance of the radicle or haustorium of the parasite seed may be regarded as "germination" in laboratory tests, it is thought likely that in presence of the host plant root the development of the haustorium is influenced by growth substances other than the stimulants. *In vitro* germination of *Striga* and *Orobanche* by purified factors brings about the production of a haustorium some 1–3 mm in length, but development beyond this point does not occur. No root hair development, nor shoot formation, is observed. At this stage it is necessary for other, post-germination, factors to be present, and these are doubtless produced by the living stimulating-plant root. It is at this point perhaps that the distinction between host and non-host is made. Sunderland (1960a,b) observed a synergistic effect on the percentage of germination of parasite seed when mixtures of linseed and maize root exudates were employed. It seems likely that this was due to a more favourable composition of post-germination factors being present in the mixed solution.

Williams (1961) has shown that application of kinetin and gibberellic acid to germinated *Striga* seed leads to development of the haustorium *in vitro*. Gibberellic acid in concentrations of 1–25 ppm promoted shoot formation, and kinetin led to development of root hairs. These effects were identical to those observed in a developing shoot near to a host plant. It appears therefore that in addition to nutrients, development of the parasite seedling requires growth hormones at an early stage. In the case of *Orobanche*, the typical result of attachment to a host plant is nodule formation. It is difficult to distinguish in these nodules the line of demarcation between *Orobanche* and host-plant cells. It is probable that a chemical factor is present in the developing seed which brings about a breakdown of the outer layers of the host-plant root and stimulates cell division in the host tissue. This would account also for the fact that where host plant roots come into contact with the rootlets arising from the more mature *Orobanche* nodule, secondary fusions readily occur, leading ultimately to a tangled mass of host and parasite roots.

III. CONCLUSION

Although the processes involved in parasitic attachment to host plants clearly involve several stages, it is likely that pre-germination conditioning and post-germination development are controlled by factors similar to those already known and partly understood. The triggering of seed germination (and possibly the attachment of haustoria to the host root) involves unique chemical factors which are just beginning to be investigated.

REFERENCES

Anderson, L. C. (1968). *Phytomorphology* 18 (2), 166.
Agricultural Research Council. Weed Research Organisation Bibliographies Nos. 1, 2, 23. (Weed Research Organisation, Begbroke Hill, Yarnton, Oxford, England.)
Atanas, M. (1969). *Bulg. Tyntyun* 14 (2), 4.
Barcinsky, R. M. (1934). *Dokl. Akad. Nauk* 1, 343.
Barcinsky, R. M. (1935). *Dokl. Akad. Nauk* 2, 311.
Beck-Managetta, G. (1930). *In* "Das Pflanzenreich" (A. Engler, ed.), Bd. 96, IV, 261, p. 348.
Bogdanova, V. M. (1969). *Bot. Zh.* (*Leningrad*) 54 (10), 1599.
Botha, P. J. (1951). *Jl S. Afr. Bot.* 17, 49.
Brian, P. W., Hemming, H. G. and Lowe, D. (1962). *Nature, Lond.* 193, 946.
Bridel, M. and Charaux, C. (1924). *Bull. Soc. Chim.* IV, 1153.
Brown, R. (1946). *Nature, Lond.* 157, 64.
Brown, R., Johnson, A. W. and Robinson, E. (1949a). *Nature, Lond.* 163, 842.
Brown, R., Johnson, A. W. and Robinson, E. (1949b). *Proc. R. Soc.* B 136, 577.
Brown, R., Johnson, A. W., Robinson, E. and Todd, A. R. (1949c). *Proc. R. Soc.* B 136, 1.
Brown, R., Greenwood, A. D., Johnson, A. W. and Long, A. G. (1951a). *Biochem. J.* 48, 559.
Brown, R., Greenwood, A. D., Johnson, A. W., Long, A. G. and Tyler, G. J. (1951b). *Biochem. J.* 48, 564.

Brown, R., Greenwood, A. D., Lansdowne, A. R., Long, A. G. and Sunderland, N. (1952). *Biochem. J.* **52**, 571.
Cappelletti, C. (1936). *Nuovo Giorn. Bot. Ital.* **43**, 263.
Chabrolin, C. (1937–1938). *Ann. Service Bot. Agron. Tunis*, **14/15**, 91–144.
Chabrolin, C. (1938). *C. R. Acad. Sci., Paris* **206**, 1990.
Chadha, K. L. and Manon, V. N. (1969). *J. Res. Punjab Agric. Univ.* **6** (3), 821.
Cook, C. E., Whichard, L. P., Turner, B., Wall, M. E. and Egley, G. H. (1966). *Science, N. Y.* **154**, 1189.
Cross, B. E. (1954). *J. chem. Soc.* 4670.
Degen, A. (1911). *Tanulmanyok az arankarol. Kiserletugyi Kozlemenyek* **14**, 493.
Dehay, C. and Care, M. (1957). *C. R. Acad. Sci., Paris*, **247**, 336.
Donini, B. (1959). *Agric. ital.* **59** (6), 219–222.
Fritsche, E., Bouillenne-Walrand, M. and Bouillenne, R., (1958). *Bull. Classe Sci. Ac. Roy. Belg.*, 5ᵉ ser., **44** (3), 613.
Griffiths, G. M., MacWilliam, I. C. and Reynolds, T. (1964). *Nature, Lond.* **202**, 1027.
Heinricher, E. (1898). *Ber. dtsch. bot. Ges.* **16**, 2.
Izard, C. (1958). *Ann. Inst. Expl Tabac. Bergerac* **3** (1), 77.
Izard, C. and Hitier, H. (1954). *Ann. Inst. Expl Tabac. Bergerac* **2**, 9.
Izard, C. and Hitier, H. (1958). *C. R. Acad. Sci., Paris* **246**, 2659.
Jones, K. Wilson (1953). *Nature, Lond.* **172**, 128.
Kadry, A. and Tewfik, H. (1956). *Svensk. Bot. Tidskr.* **50**, 270.
Klingman, G. C. and Moreland, D. E. (1962). *Nature, Lond.* **195**, 199.
Krenner, J. A. (1958). *Acta Bot. Acad. Sci. Hung.* **4** (1/2), 113.
Lansdowne, A. R. (1953). Ph.D. Thesis, Cambridge University.
Long, A. G. (1953). Ph.D. Thesis, Cambridge University.
Muller, K. O. (1941). *Phytopath. Z.* **13**, 530.
Nash, S. M. and Wilhelm, S. (1960). *Phytopathol.* **50** (10), 772.
Netzer, D. (1961). *Hassadeh* **41** (12), 1515.
Netzer, D. (1969). *Plant Dis. Rep.* **53** (9), 731.
Parker, C. (1965). *Pest Articles and News Summaries*, Section C, **11**, 99.
Passalacqua, T. (1934). *Ann. Tech. Agr.* **7**, 487.
Pearson, W. H. H. (1912). *Agric. J. Un. S. Afr.* **3**, 651.
Privat, G. (1960). *Ann. Sci. Nat. Bot.* **12**, 721.
Puzzilli, M. (1969). *Tobacco* **73** (730), 6.
Racovitza, A. (1958). *J. Agric. Trop. Bot. Appl.* **6–7**, 111.
Rao, P. G. (1955). *Sci. Cult. (Calcutta)* **21** (5), 258.
Sembdner, G. (1963). *Physiol. Oekol. Biochem. Keimung. Mater. Int. Symp.* **1**, 239.
Sunderland, N. (1960a). *J. exp. Bot.* **11** (32), 236.
Sunderland, N. (1960b). *J. exp. Bot.* **11** (33), 356.
Thompson, P. A. (1969). *Hort. Rev.* **9** (2), 130.
Tyler, G. J. (1952). Ph.D. Thesis, Cambridge University.
Villiers, T. A., Frankland, B., and Wareing, P. F. (1963). Physiol. Oekol. Biochem. Keimung. Mater. Int. Sympt., **1**, 301.
Wilhelm, S. (1962). XVIth International Horticultural Congress (Brussels).
Wilhelm, S., Storkan, R. C., Sagan, J. E. and Carpenter, T. (1959). *Phytopathol.* **49**, 530.
Williams, C. N. (1961). *W. Afr. Jl Biol. Chem.* **2**, 57.
Williams, C. (1961). *Nature, Lond.* **189**, 378.
Worsham, A. D., Moreland, D. E. and Klingman, G. C. (1964). *J. exp. Bot.* **15** (45), 556.
Zahran, M. K. (1970). *Proc. 10th Br. Weed Control. Conf.* **2**, 680.

Author Index

Numbers in italics are those pages on which references are listed

10*

Gunsalus, I. C., 156, *161*
Günther, W. H., 148, 149, *160*
Gupta, V. K., 133, *143*
Guthrie, F. E., 2, *12*
Guyot, H., 104, *123*
Gyorgyi, P., 183, *199*

H

Haase, E., 4, 9, *11*
Hadwiger, L. A., 221, *233*
Hahlbrock, K., 107, 108, *123*
Hahn, G. A., 150, *160*
Haisman, D. R., 106, *123*
Hall, H. H., 132, 141, *143*, *144*
Hallinan, E. A., 170, *175*
Halpern, B. P., 97, *101*
Halver, J. E., 133, *143*
Halvorson, H. O., 106, *122*
Hamasaki, T., 136, *143*
Hanawalt, R. B., 204, *216*
Hankes, L. V., 113, *124*
Hansen, R. E., 156, *161*
Hanson, F. E., 22, *23*
Harborne, J. B., 168, *177*
Hardegger, E., 219, *232*
Harman, R. R. M., 5, *12*
Harp, A. R., 130, *143*
Harper, J. S., 169, *176*
Harper, R., 46, *56*, 62, *69*
Harrison, J. S., 169, *176*
Hatsuda, Y., 136, *143*
Havran, R. T., 158, *160*
Hawkes, J. G., 47, *56*
Head, M. A., 184, *200*
Heady, H. F., 72, *100*
Heath, D., 183, *199*
Heathcote, J. G., 134, *143*
Hegarty, M. P., 171, 172, *175*, *176*
Hegnauer, R., 104, 118, *123*
Heinricher, E., 239, *248*
Hemming, H. G., 244, *247*
Henderson, J. H. M., 91, *100*
Hendrickson, H. R., 108, 109, 119, *122*
Henery-Logan, K. P., 135, *144*
Henke, L. A., 171, *176*
Hennig, E., 27, 28, 29, *41*, *42*
Henning, G. J., 60, *69*
Henry, T. A., 110, 119, *122*, *123*
Henson, L., 93, *100*
Herissey, H., 112, *123*
Hesseltine, C. W., 130, 133, *142*, *143*
Hewson, P. R., 90, *101*

Higgins, V. J., 112, *124*, 224, *232*, *233*
Himwich, W. A., 111, *123*
Hironaka, R., 77, 89, *100*
Hirsch, M. L., 150, *161*
Hirst, E. L., 95, *100*
Hitier, H., 237, 238, 240, *248*
Hiura, M., 222, *232*
Ho, P., 165, *176*
Hocking, B., 4, *11*
Hodges, F. A., 130, *143*
Hodges, R., 137, *142*
Hodgson, E., 2, *12*
Hoffman, J. L., 149, 150, 151, *160*
Hohl, H. R., 219, *232*
Holgate, M. D., 91, *100*
Holker, J. S. E., 136, 137, *143*
Holzapfel, C. W., 134, *143*
Honkanen, E., 94, 97, *100*
Hook, D. D., 211, *215*
Hopkins, G. H. E., 5, *11*
Horowitz, N. H., 169, *175*
Horowitz, R. M., 98, *100*
Hoskins, F. A., 96, 97, *101*
House, H. L., 13, *23*
Hsiao, T. H., 3, *11*, 21, *23*
Huber, J., 14, *23*
Huber, R. E., 154, 155, 157, 158, *160*, *161*
Hughes, R. D., 27, 32, 36, *43*
Hurd-Karrer, A. M., 146, 147, *160*
Hurst, J. J., 106, *123*
Hutton, E. M., 106, *123*, 171, *175*
Hylin, J. W., 171, 172, *175*

I

Ikawa, M., 168, *176*
Imaizumi, K., 223, *232*
Ingham, J. L., 230, *232*
Insalata, N. F., *142*
Iongh, M. de, 133, *143*
Irvine, F. R., 170, *176*
Ishizaka, N., 223, 231, *233*
Iyer, V. N., 198, *199*
Izard, C., 237, 238, 240, *248*

J

Jaag, O., 219, *232*
Jager, H., 7, *11*
Jago, M. V., 183, 186, 187, 189, 190, 191, 195, 196, 197, 198, *199*
Janzen, D. H., 169, 170, *175*

Index of Plant Species

Index of Animal Species

A

Acraea, 9
A. encedon, 9
Acromyrmex, 14
A. octospinosus, 14–18, 20, 21
Acyrthosiphon pisum, 29–31
Agriolimax caruanae, 117
A. reticulatus, 116, 117
Amauris, 5
Amphicallia bellatrix, 7
Apanteles tetricus, 111
A. zygaenarum, 111
Aphis fabae, 27, 28, 33
A. nerii, 4, 7
Arianta arbustorum, 117
Arion ater, 117
A. hortensis, 117
A. subfuscus, 117
Arctia caja, 5, 7, 8
Atta, 14
A. capiguara, 15
A. cephalotes, 14–22
A. colombica tonsipes, 14
A. laevigata, 15
A. texana, 14
Aularches miliaris, 4

B

Brevicoryne brassicae, 26, 33–37, 39–41

C

Caenocoris nerii, 7
Cepaea hortensis, 117
C. nemoralis, 117
Coccinella septempunctata, 3
Coturnix japonicus, 3

D

Danaus chrysippus, 8, 9
D. plexippus, 7

G

Gazella gazella, 3

H

Helicella virgata, 117
Heliothis zea, 22
Helix aspera, 117, 118
H. hespera, 116
Hypera plantaginis, 7, 111

L

Labidomera clavicollis, 3
Leptinotarsa decimlineata, 3
Linyphia triangularis, 22

M

Manduca sexta, 22
Maniola jurtina, 111
Megoura viciae, 28
Milax budapestensis, 117
Monacha cartusiana, 117
Myzus ascalonicus, 26, 31, 33
M. persicae, 28, 33–41

O

Operophtera brumata, 21

P

Papilio dardanus, 8
Parasemia plantaginis, 8
Pemphigus bursarius, 26, 33
Pieris rapae, 32
Poekilocerus, 4
P. bufonius, 6, 7
Polyommatus icarus, 7, 111
Procavia, 3

265

Subject Index

(Where a page number is in italics, the compound referred to is first shown structurally)

A

Acyrthosiphon pisum, activity on host and non host plants, 30
Adenostoma fasciculatum, toxins from, 205–210
Aflatoxins, 127–141
  carcinogenic effects of, 133
  fluorescence of, 130, 131
  history of research on, 129, 130
  interaction with rubratoxins, 141, 142
  structure of, *131*
  suggested biosynthesis of, 137–140
  toxicity of, 132
Alkaloid metabolites, as causes of toxicity, 184, 185
Alkaloids,
  feeding response of ruminants to, 91, 92
  of *Senecio*, 179–199
Allylisothiocyanate, *34*
  relation of aphid growth to concentration of, 40
Amino acid precursors, of cyanogenic glucosides, 107
Amino acids,
  attraction of aphids to, 35
  relation of aphid growth to concentration of, 40
Amino acids, toxic,
  in *Lathyrus*, 167
  in the Leguminosae, 163–174
α-Amino-γ-oxalylaminobutyric acid, structure and effects of, 166
α-Amino-β-oxalylaminopropionic acid, structure and effects of, 165
Amygdalin, *53*
Anthraquinones, from *Aspergillus versicolor*, 136, 137
Antimicrobial compounds,
  production in response to cell damage, 230
  universal presence of, 231

Ants,
  foraging activity of, 19, 20
  leaf attack by, 13–22
  substrate preference of, 18–21
Aphids,
  as phytochemists, 25–41
  host distribution and specificity of, 26, 27
  host selection behaviour of, 27–34
  phloem feeding of, 28, 29
  relative growth rate of, 36–41
Arbutin, *206*
Arctostaphylos glandulosa, toxins from, 207–210
Aromatic substances, classification of, 62
Aspergillus flavus, 131
  aflatoxins from, 134, 135
Aspergillus versicolor, metabolites from, 135–137
Astringency, causes of, 53
Azetidine-2-carboxylic acid, structure, occurrence and effects of, 172, 173

B

Bacterial diseases,
  and production of antibacterial substances, 230
  effect on phaseollin, 229
Beans, disease resistance in, 225–231
Behavioural responses,
  by ants to fungus substrates, 17–21
  of ruminants to chemical solutions, 74
Birds, effect on life patterns of insects, 2, 3
Bitterness,
  as universally repellent character, 49
  causes of, 53
Brevicoryne brassicae, feeding habits of, 34–41

V

Vanillin, *63*
Vicia faba,
 infection of, by *Botrytis*, 225
 inhibition of *Botrytis* by phytoalexin
 in, 225, 226

W

Wyerone, *226*
Wyerone acid, *226*
 produced by *Vicia faba* in response to
 infection, 226–228

X

Xanthocillin X, *131*
Xanthones, from *Aspergillus versicolor*,
 135
D-Xylulose, *240*
 as germination stimulant, 240

Z

Zingerone, *63*
Zingerone-related compounds, *64*, 65
 pungency of, 65